FUNDAMENTALS OF
CONTINUUM MECHANICS

FUNDAMENTALS OF CONTINUUM MECHANICS

John W. Rudnicki
Northwestern University, USA

WILEY

Reprinted with corrections May 2015

This edition first published 2015
© 2015 John Wiley & Sons, Ltd

Registered office
John Wiley & Sons Ltd, The Atrium, Southern Gate, Chichester, West Sussex, PO19 8SQ, United Kingdom

For details of our global editorial offices, for customer services and for information about how to apply for permission to reuse the copyright material in this book please see our website at www.wiley.com.

Library of Congress Cataloging-in-Publication Data

Rudnicki, John W.
    Fundamentals of continuum mechanics / John W. Rudnicki.
      pages cm
    Includes bibliographical references and index.
    ISBN 978-1-118-47991-9 (paperback)
  1. Continuum mechanics.   I. Title.
    QA808.2.R83 2015
    531.01′51–dc23

                                    2014032058

Set in 10/12pt Times by Aptara Inc., New Delhi, India

1   2015

To Ellen

Contents

Preface

The first question anyone contemplating writing a book on continuum mechanics must ask themselves is "Why?" There are numerous continuum mechanics books and many of them are very good. I have learned from them and I have used some of them in teaching a course on continuum mechanics. Yet, none seemed entirely suitable for the particular course that I have taught for many years at Northwestern University. The course is one quarter, 10 weeks, typically taught to first-year graduate students. Many of the students are PhD students in fluid, solid, or structural mechanics. For them the course is meant to be a foundation for more advanced and specialized topics. But there are also PhD students from geotechnical and biomedical engineering, materials science, geology, and geophysics. In recent years, the course has attracted many Master's degree students who do not intend to go on for the PhD. For these groups of students, the course is likely to be their only exposure to much of this material.

To satisfy this diverse audience within a 10-week quarter, the course must be concise. It must begin at a level suitable for students without a strong background in mechanics yet be sufficiently general and advanced to be a solid foundation for students continuing in mechanics. Most continuum mechanics books include chapters on elasticity, fluid mechanics, viscoelasticity, and other applications to particular material behaviors. In a 10-week quarter there is time to treat only the barest minimum of this material. At Northwestern and, I am sure, at many other universities, students go on to take courses devoted to these subjects and there are good books devoted to them. On the other hand, with the development of computational mechanics and advanced materials, the diversity and complexity of constitutive models that may be encountered in research is immense and growing. Although this diversity makes a thorough treatment of the subject impossible in one quarter, it makes a solid background in the fundamentals of continuum mechanics even more essential.

These requirements might seem to relegate a book based on the course to a niche. Yet I became convinced that there is a need for such a book not only for students but also for practitioners. I came to this opinion through the encouragement of many students who used the notes on which the book is based and feedback from colleagues at other universities. I view the book as the "baby bear" of continuum mechanics books: neither too long nor too short, neither too advanced nor too elementary, neither too superficial nor too in-depth. Although it is based on a one-quarter course, it would be suitable for a one-semester course by expanding the treatment of some material. Possibilities are suggested by some of the examples and exercises. In teaching the course and writing the book I do not attempt to be mathematically rigorous. Instead, I try to emphasize operational definitions and doing "what comes naturally" as suggested by the notation. I believe that the notation can be an important aide for both beginning and advanced students to gain confidence and as a foundation for more in-depth understanding.

The goals of the course are a basic understanding of the following: tensors and tensor calculus in Cartesian coordinate systems and in coordinate-free form; stress as a tensor; the difference between material and spatial descriptions of motion; the measures of strain and deformation and deformation rate for arbitrary deformation magnitudes; the formulation of equations describing balance of mass, momentum, and energy in their various forms; and the introduction of constitutive behavior.

After reading the book, students should be familiar with and be able to do the following: use index notation for vectors and tensors; calculate components of tensors in different Cartesian coordinate systems; use and manipulate the stress tensor and explain the meaning of its components; manipulate and describe the relations among different measures of strain and deformation; derive and explain equations describing the balance of mass, momentum and energy; and, perhaps most importantly, read and understand papers and texts on advanced continuum mechanics.

The book is divided into five parts: Mathematical Preliminaries; Stress; Motion, and Deformation; Balance of Mass, Momentum, and Energy; and Ideal Constitutive Relations. Although the first part is preliminary, it comprises nearly a third of the length. It introduces notation and forms a foundation for the remainder of the book. Many of the exercises derive results that are used later in the book. In teaching the course, I resist the urge to move rapidly through this material. Although some of the material is elementary the approach is likely to be new even to readers who have some familiarity with these topics.

One of the most pleasurable tasks of writing a book is to thank the many people who provided help and support, although I cannot mention them all by name. As usual, it is necessary to emphasize that none of them are responsible for any shortcomings of the book in fact, concept, or clarity. First of all, I want to thank my daughter Jean who suffered working for her father one summer to translate my handwritten notes into LaTex and to prepare the first drafts of figures. I am indebted to my teachers at Brown University, especially Professor Jacques Duffy who taught me my first course in continuum mechanics, Professor Ben Freund, and Professor James Rice (now at Harvard) who was my undergraduate and PhD advisor and continues to teach me mechanics every time I see him. The students who took my class exposed my incomplete understanding, improved the clarity of the exposition, and provided both encouragement and criticism. Steve Sun and Miguel Bessa read the draft and provided corrections and helpful comments. The continuum mechanics class in the Fall of 2013 responded enthusiastically to my offer to earn points by finding misprints, errors, and unclear passages. I especially want to thank Aaron Stebner of the Colorado School of Mines who used the draft for a course on continuum mechanics and provided invaluable feedback on the text and problems. Tom Carter, my editor, was unfailingly helpful, encouraging, and responsive. I am grateful to Northwestern University for providing a stimulating environment for more than 30 years and for a leave of absence during which I did much of the preparation of the book. Finally, I do not have words to express thanks for the support of my family in everything I do.

Nomenclature

Notation:

Bold-faced upper case letters refer to (mainly second-order) tensors.

Bold-faced lower case letters refer to vectors. The magnitude of a vector is denoted by the same letter in italics or $|\ldots|$.

Subscripts are denoted i through v and range over 1, 2, and 3. Upper case subscripts and Roman numerals refer to principal values.

Greek subscripts α, β, etc., range over 1 and 2.

α	thermal diffusivity
α, β	scalars
γ	shear
δ_{ij}	Kronecker delta
ϵ_{ijk}	permutation symbol
ε	infinitesimal (small) strain
η	dynamic viscosity
θ	temperature
Θ	bulk viscosity
κ	thermal conductivity tensor
λ	principal value (eigenvalue)
λ, μ	Lamé constants
Λ	stretch ratio
μ	friction coefficient
μ	shear modulus
μ	shear viscosity
$\boldsymbol{\mu}$	unit vector in principal direction (eigenvector)
v	Poisson's ratio
ξ	non-dimensional length
ρ	mass density in current configuration
ρ_0	mass density in reference configuration
σ	Cauchy stress
σ'	deviatoric Cauchy stress
$\hat{\sigma}$	rotationally invariant Cauchy stress
$\bar{\sigma}_{ij}$	Cauchy stress in the reference configuration
τ	Kirchhoff stress

$\boldsymbol{\phi}(\mathbf{X}, t)$	motion
$\boldsymbol{\Omega}$	infinitesimal rotation tensor
a	area in current configuration
A	area in reference configuration
$\mathbf{a}(\mathbf{x}, t)$	acceleration, Eulerian description
$\mathbf{A}(\mathbf{X}, t)$	acceleration, Lagrangian description
$\mathbf{A}$	orthogonal tensor
$\mathcal{A}(\mathbf{x}, t)$	a vector or scalar quantity proportional to the mass
$\mathbf{b}$	body force per unit mass
$\mathbf{b}^0$	body force per unit mass in the reference configuration
$\mathbf{B}$	left Cauchy–Green tensor
$\mathbf{B}^{-1}$	Cauchy deformation tensor
$\mathbf{B}, \mathbf{B}^{-1}$	Finger tensors
c_L, c_s	bulk (dilatational) and shear wave speeds
c_p, c_v	specific heats at constant pressure and volume
c_{pq}	cofactor
$\mathbf{C}$	Green or right Cauchy–Green deformation tensor
C_{ijkl}	components of elastic modulus tensor
$\mathbf{D}$	rate of deformation tensor
E	total energy
E	Young's modulus
$\mathbf{e}_1, \mathbf{e}_2, \mathbf{e}_3$	orthonormal base vectors
$\mathbf{e}^A$	Almansi strain
$\mathbf{E}$	material strain tensor
$\mathbf{E}^G$	Green–Lagrange strain tensor
$\mathbf{E}^{(1)}$	Biot strain tensor
$\mathbf{E}^{(\ln)}$	logarithmic strain tensor
$f(\Lambda)$	scale function for material strain
$\mathbf{F}$	deformation gradient tensor
$g(\lambda)$	scale function for spatial strain
h	height, thickness
$\mathbf{H}^{-1}$	inverse of tensor $\mathbf{H}$
$\mathbf{I}$	identity tensor
I_1, I_2, I_2	principal invariants
IE	internal energy
J	$\lvert \partial \mathbf{x} / \partial \mathbf{X} \rvert = \lvert \partial x_i / \partial X_j \rvert$
K	bulk modulus
KE	kinetic energy
$\mathbf{L}$	velocity gradient tensor
m	mass
M	matrix
M_{ij}	components of a matrix
M_{li}^*	components of the adjugate
$\mathbf{n}$	unit normal in current configuration
$\mathbf{N}$	unit normal in reference configuration
p	pressure

P	power input
$\mathbf{q}$	heat flux per unit current area
$\mathbf{Q}$	heat flux per unit reference area
$\dot{Q}$	rate of heat input
$\mathbf{Q}(t)$	rigid body rotation
r	radial coordinate
r	rate of internal heating (heat source) per unit mass in current state
R	rate of internal heating (heat source) per unit mass in reference state
$\mathbf{R}$	rotation tensor in polar decomposition
$\mathbf{S}$	work-conjugate stress tensor
$\mathbf{S}^{PK2}$	second Piola–Kirchhoff stress
$\mathbf{t}$	traction, surface force per unit current area
$\mathbf{t}^0$	nominal traction, surface force per unit reference area
$\mathbf{T}^0$	nominal (first Piola–Kirchhoff) stress
u	internal energy per unit mass
$\mathbf{u}$	displacement
$\mathbf{U}, \mathbf{V}$	deformation tensors in polar decomposition
v	specific volume
v	volume in current configuration
V	volume in reference configuration
V_{ijkl}	components of constitutive tensor for Newtonian fluid
$\mathbf{v}(\mathbf{x}, t)$	velocity, Eulerian description
$\mathbf{V}(\mathbf{X}, t)$	velocity, Lagrangian description
W	internal energy per unit reference volume
$\mathbf{w}$	dual or polar vector
$\mathbf{W}$	vorticity or spin tensor
$\dot{W}_0$	rate of stress working per unit reference volume
$\mathbf{x}$ or x_1, x_2, x_3	position of a material particle in the current configuration
$\mathbf{X}$ or X_1, X_2, X_3	position of a material particle in the reference configuration
$\boldsymbol{\nabla}$	gradient operator
$[\ldots]$	matrices, column or row vectors
$\lvert \ldots \rvert$	determinant or magnitude

Introduction

Continuum mechanics is a mathematical framework for studying the transmission of force through and deformation of materials of all types. The goal is to construct a framework that is free of special assumptions about the type of material, the size of deformations, the geometry of the problem, and so forth. Of course, no real materials are actually continuous. We know from physics and chemistry that all materials are formed of discrete atoms and molecules. Even at much larger size scales, materials may be composed of distinct components, e.g., grains of sand or platelets of blood. At even larger scales, for instance, the Earth's crust, fractures are ubiquitous. Nevertheless, treating material as continuous is a great advantage because it allows us to use the mathematical tools of continuous functions, such as differentiation. In addition to being convenient, this approach works remarkably well. This is true even at size scales for which the justification of treating the material as a continuum might be debatable. Although there are certainly problems for which it is necessary to take into account the discrete nature of materials, the ultimate justification for using continuum mechanics is that predictions are often in accord with observations and measurements.

Although the framework of continuum mechanics does not make reference to particular kinds of materials, its application does require a mathematical description of material response. These descriptions are inevitably idealizations based on experiments, conceptual models, or microstructural considerations.

Until recently, it was only possible to solve a relatively small number of problems without the assumptions of small deformations and very simple material behavior. Now, however, modern computational techniques have made it possible to solve problems involving large deformation and complex material behavior. This possibility has made it important to formulate these problems correctly and to be able to interpret the solutions. Continuum mechanics does this.

The vocabulary of continuum mechanics involves mathematical objects called tensors. These can be thought of as following naturally from vectors. Therefore, we will begin by studying vectors. Although most students are acquainted with vectors in some form or another, we will reintroduce them in a way that leads naturally to tensors.

Fundamentals of Continuum Mechanics, First Edition. John W. Rudnicki.
© 2015 John Wiley & Sons, Ltd. Published 2015 by John Wiley & Sons, Ltd.

Part One

Mathematical Preliminaries

This part provides the foundation for the rest of the book. The treatment is meant to make the book self-contained, assumes little background from the reader, and only covers what is needed later in the book. The treatment begins with vectors. Although most readers will be acquainted with vectors, they are introduced in a way that leads naturally to tensors, introduced in the second chapter, and their representation in terms of dyadics, in the third. Vectors and tensors are introduced in coordinate-free form, appropriate for describing the physical entities that arise in continuum mechanics, before discussing their representation in terms of Cartesian coordinates in the third chapter. This chapter introduces index notation and the summation convention. Chapter 4 discusses the cross product, introduces the permutation symbol, and provides an introduction to the discussion of determinants in the following chapter. Chapter 6 derives the relation between vector and tensor components in coordinate systems that differ by a rotation. This relation provides an alternative method of defining vectors and tensors. Chapter 7 discusses principal values and directions which are pertinent to many of the particular tensors introduced later. Chapter 8 discusses the gradient, but this material is not needed until Part Three, Motion and Deformation, and can be deferred until then. Although Chapter 18, Transformation of Integrals, covers a subject more naturally suited to this part, it is not needed until Part Four and is deferred until then.

Part One has been written with a view toward what is used later in the book. Many of the exercises derive results that are used later in the book. Consequently, even readers who are familiar with much of this material may find value here.

Fundamentals of Continuum Mechanics, First Edition. John W. Rudnicki.
© 2015 John Wiley & Sons, Ltd. Published 2015 by John Wiley & Sons, Ltd.

1

Vectors

Some physical quantities are described by scalars, e.g., density, temperature, kinetic energy. These are pure numbers, although they do have dimensions. It would make no physical sense to add a density, with dimensions of mass divided by length cubed, to kinetic energy, with dimensions of mass times length squared divided by time squared.

Vectors are mathematical objects that are associated with both a magnitude, described by a number, and a direction. An important property of vectors is that they can be used to represent physical entities such as force, momentum, and displacement. Consequently, the meaning of the vector is (in a sense we will make more precise) independent of how it is represented. For example, if someone punches you in the nose, this is a physical action that could be described by a force vector. The physical action and its result (a sore nose) are independent of the particular coordinate system we use to represent the force vector. Hence, the meaning of the vector is not tied to any particular coordinate system or description. For this reason, we will introduce vectors in *coordinate-free* form and defer description in terms of particular coordinate systems.

A vector **u** can be represented as a directed line segment, as shown in Figure 1.1. The length of the vector is its magnitude, and denoted by u or by $|\mathbf{u}|$. Multiplying a vector by a positive scalar α changes the length of the vector but not its orientation. If $\alpha > 1$, the vector $\alpha\mathbf{u}$ is longer than **u**; if $\alpha < 1$, $\alpha\mathbf{u}$ is shorter than **u**. If α is negative, the orientation of the vector is reversed. It is always possible to form a vector of unit magnitude by choosing $\alpha = u^{-1}$.

The addition of two vectors **u** and **v** can be written as

$$\mathbf{w} = \mathbf{u} + \mathbf{v} \qquad (1.1)$$

Although the same symbol is used as for ordinary addition, the meaning here is different. Vectors add according to the parallelogram law shown in Figure 1.2. If the "tails" of the vectors (the ends without arrows) are placed at a point, the sum is the diagonal of the parallelogram with sides formed by the vectors. Alternatively the vectors can be added by placing the "tail" of one at the "head" of the other. The sum is then the vector directed from the free "tail" to the free "head." Implicit in both of these operations is the idea that we are dealing with "free"

Fundamentals of Continuum Mechanics, First Edition. John W. Rudnicki.
© 2015 John Wiley & Sons, Ltd. Published 2015 by John Wiley & Sons, Ltd.

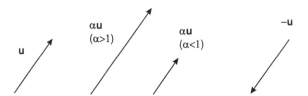

Figure 1.1 Multiplication of a vector by a scalar.

vectors. In order to add two vectors, they can be moved, keeping the length and orientation, so that the vectors can be connected head to tail. It is clear from the construction in Figure 1.2 that vector addition is commutative:

$$\mathbf{w} = \mathbf{u} + \mathbf{v} = \mathbf{v} + \mathbf{u}$$

Note the importance of distinguishing vectors from scalars; without the bold face denoting vectors, equation (1.1) would be incorrect: the magnitude of $\mathbf{w}$ is not the sum of the magnitudes of $\mathbf{u}$ and $\mathbf{v}$.

The parallelogram rule for vector addition follows from the nature of the physical quantities, e.g., velocity and force, that vectors represent. The rule for addition is an essential element of the definition of a vector that can distinguish them from other quantities that have both length and direction. For example, finite rotations about three orthogonal axes can be characterized by length and magnitude. Finite rotation cannot, however, be a vector because addition is not commutative. To see this, take a book with its front cover up and binding to the left. Looking down on the book, rotate it 90° counterclockwise. Now rotate the book 90° about a horizontal axis counterclockwise looking from the right. The binding should be on the bottom. Performing these two rotations in reverse order will orient the binding toward you.

Hoffmann (1975) relates the story of a tribe that thought spears were vectors because they had length and magnitude. To kill a deer to the northeast, they would throw two spears, one to the north and one to the east, depending on the resultant to strike the deer. Not surprisingly, there is no trace of this tribe, which only confirms the adage that "a little knowledge can be a dangerous thing."

The procedure for vector subtraction follows from multiplication by a scalar and addition. To subtract $\mathbf{v}$ from $\mathbf{u}$, first multiply $\mathbf{v}$ by -1, then add $-\mathbf{v}$ to $\mathbf{u}$:

$$\mathbf{w} = \mathbf{u} - \mathbf{v} = \mathbf{u} + (-\mathbf{v})$$

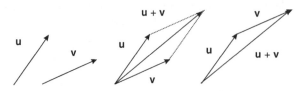

Figure 1.2 Addition of two vectors.

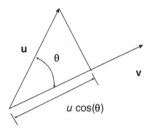

Figure 1.3   Scalar product.

There are two ways to multiply vectors: the scalar or dot product and the vector or cross product. The scalar product is given by

$$\mathbf{u} \cdot \mathbf{v} = uv\cos(\theta) \tag{1.2}$$

where θ is the angle between $\mathbf{u}$ and $\mathbf{v}$. As indicated by the name, the result of this operation is a scalar. As shown in Figure 1.3, the scalar product is the magnitude of $\mathbf{v}$ multiplied by the projection of $\mathbf{u}$ onto $\mathbf{v}$, or vice versa. The definition (1.2) combined with rules for vector addition and multiplication of a vector by a scalar yield the relation

$$(\alpha\mathbf{u}_1 + \beta\mathbf{u}_2){\cdot}\mathbf{v} = \alpha\mathbf{u}_1{\cdot}\mathbf{v} + \beta\mathbf{u}_2{\cdot}\mathbf{v}$$

where α and β are scalars and $\mathbf{u}_1$ and $\mathbf{u}_2$ are vectors.

If $\theta = \pi$ in (1.2) the two vectors are opposite in sense, i.e., their arrows point in opposite directions. If $\theta = \pi/2$ or $-\pi/2$, the scalar product is zero and the two vectors are *orthogonal*. Although the scalar product is zero neither $\mathbf{u}$ nor $\mathbf{v}$ is zero. If, however,

$$\mathbf{u} \cdot \mathbf{v} = 0 \tag{1.3}$$

for any vector $\mathbf{v}$ then $\mathbf{u} = 0$.

The other way to multiply vectors is the vector or cross product. The result is a vector

$$\mathbf{w} = \mathbf{u} \times \mathbf{v} \tag{1.4}$$

The magnitude is $w = uv\sin(\theta)$, where θ is again the angle between $\mathbf{u}$ and $\mathbf{v}$. As shown in Figure 1.4, the magnitude of the cross product is equal to the area of the parallelogram

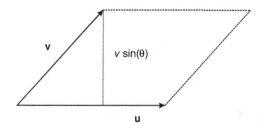

Figure 1.4   Magnitude of the vector or cross product.

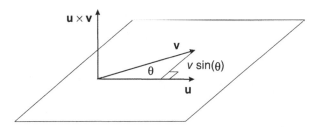

Figure 1.5   Direction of vector or cross product.

formed by **u** and **v**. As depicted in Figure 1.5, the direction of **w** is perpendicular to the plane formed by **u** and **v** and the sense is given by the *right hand rule*: If the fingers of the right hand are in the direction of **u** and then curled in the direction of **v**, the thumb of the right hand is in the direction of **w**. The three vectors **u**, **v**, and **w** are said to form a right-handed system.

The triple scalar product $(\mathbf{u} \times \mathbf{v}) \cdot \mathbf{w}$ is equal to the volume of the parallelepiped formed by **u**, **v**, and **w** if they are right-handed and the negative of the volume if they are not (Figure 1.6). The parentheses in this expression may be omitted because it makes no sense if the dot product is taken first: the result is a scalar and the cross product is an operation between two vectors.

Now consider the triple vector product $\mathbf{u} \times (\mathbf{v} \times \mathbf{w})$. The vector $\mathbf{v} \times \mathbf{w}$ must be perpendicular to the plane containing **v** and **w**. Hence, the vector product of $\mathbf{v} \times \mathbf{w}$ with another vector **u** must result in a vector that is in the plane of **v** and **w**. Consequently, the result of this operation may be represented as

$$\mathbf{u} \times (\mathbf{v} \times \mathbf{w}) = \alpha \mathbf{v} + \beta \mathbf{w} \tag{1.5}$$

where α and β are scalars.

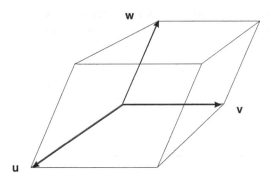

Figure 1.6   Triple scalar product.

1.1 Examples

1.1.1

Show that if the triple scalar product vanishes

$$\mathbf{u} \times \mathbf{v} \cdot \mathbf{w} = 0 \tag{1.6}$$

the three vectors are coplanar.

The *vector* product $\mathbf{u} \times \mathbf{v}$ is perpendicular to $\mathbf{u}$ and $\mathbf{v}$. If the triple scalar product vanishes, then $\mathbf{w}$ is perpendicular to $\mathbf{u} \times \mathbf{v}$ and hence is in the plane of $\mathbf{u}$ and $\mathbf{v}$. Consequently, $\mathbf{w}$ can be expressed as a linear combination of the other two, e.g., $\mathbf{w} = \alpha\mathbf{u} + \beta\mathbf{v}$ where α and β are scalars (as long as $\mathbf{u}$ and $\mathbf{v}$ are not collinear).

1.1.2

Show that if $\mathbf{w} = \alpha\mathbf{u} + \beta\mathbf{v}$ the triple scalar product of the three vectors vanishes.

Substituting $\mathbf{w}$ into (1.6) yields zero because the scalar products of $\mathbf{u} \times \mathbf{v}$ with $\mathbf{v}$ and with $\mathbf{u}$ are zero.

Exercises

1.1 Explain (in words and/or diagrams) why

$$\mathbf{u} \times \mathbf{v} = -\mathbf{v} \times \mathbf{u}$$

and that

$$\mathbf{w} \cdot \mathbf{u} = \mathbf{w} \cdot \mathbf{v} = 0$$

where $\mathbf{w} = \mathbf{u} \times \mathbf{v}$.

1.2 Explain (in words and/or diagrams) why

$$\mathbf{u} \times \mathbf{v} \cdot \mathbf{w} = \mathbf{v} \times \mathbf{w} \cdot \mathbf{u} = \mathbf{w} \times \mathbf{u} \cdot \mathbf{v}$$

but that a minus sign is introduced if the order of any two vectors is reversed.

1.3 Explain why $\mathbf{u} \times (\mathbf{v} \times \mathbf{u})$ is orthogonal to $\mathbf{u}$ and show that α and β in (1.5) are then related by

$$\alpha v \cos(\theta) + \beta u = 0$$

where θ is the angle between $\mathbf{u}$ and $\mathbf{v}$.

1.4 Prove that if (1.3) is satisfied for *any* vector $\mathbf{v}$ then $\mathbf{u} = \mathbf{0}$.

1.5 Show that
(a) $(\mathbf{u}+\mathbf{v})\cdot(\mathbf{u}-\mathbf{v}) = u^2 - v^2$
(b) $(\mathbf{u}+\mathbf{v})\times(\mathbf{u}-\mathbf{v}) = -2\mathbf{u}\times\mathbf{v}$

1.6 Consider the plane triangle shown in Figure 1.7 with sides of lengths a, b, and c and angles α, β, and γ opposite sides a, b, and c, respectively. Use *coordinate-free vector methods* to prove (do not use geometry or, if you know it, index notation)
(a) law of cosines:

$$a^2 + b^2 - 2ab\cos(\gamma) = c^2$$

(b) law of sines:

$$\frac{a}{\sin\alpha} = \frac{b}{\sin\beta} = \frac{c}{\sin\gamma}$$

[Hint: Use scalar and vector products.]

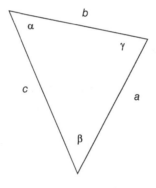

Figure 1.7   Diagram for Problem 1.6.

1.7 Let $\mathbf{a}$, $\mathbf{b}$, and $\mathbf{c}$ be non-coplanar vectors that form three edges of a tetrahedron (see Figure 1.8). Let $\mathbf{n}_1$, $\mathbf{n}_2$, and $\mathbf{n}_3$ be the outward unit normals to the faces formed by each pair of vectors and let S_1, S_2, and S_3 be the corresponding areas. Show that the product of the unit vector normal to the fourth face and the area of the face is given by

$$\mathbf{n}S = -(\mathbf{n}_1 S_1 + \mathbf{n}_2 S_2 + \mathbf{n}_3 S_3)$$

1.8 Determine α and β in (1.5) (in terms of $\mathbf{u}$, $\mathbf{v}$, $\mathbf{w}$, and scalar and cross products).

1.9 A line in direction $\mathbf{l}$ is defined by the vector relation

$$\mathbf{u} = \mathbf{a} + \mathbf{l}s$$

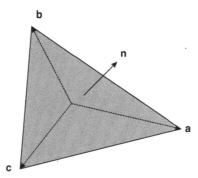

Figure 1.8   Diagram for Problem 1.7.

where **l** is a unit vector and s is a scalar parameter $-\infty < s < \infty$. Show that this will intersect a second line $\mathbf{v} = \mathbf{b} + \mathbf{m}s$, where **m** is a unit vector, if

$$\mathbf{a} \cdot (\mathbf{l} \times \mathbf{m}) = \mathbf{b} \cdot (\mathbf{l} \times \mathbf{m})$$

and determine their point of intersection, i.e. values of s for each line at the intersection.

1.10   Find the equation of the line that passes through two given points A and B located relative to a point O by two vectors **u** and **v** (Figure 1.9).

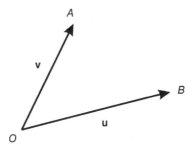

Figure 1.9   Diagram for Problem 1.10.

1.11   If **u**, **v**, and **w** are not coplanar, then it is possible to find scalars α, β, and γ such that any arbitrary vector **z** can be expressed as $\mathbf{z} = \alpha\mathbf{u} + \beta\mathbf{v} + \gamma\mathbf{w}$. Determine α, β, and γ (in terms of the vectors **u**, **v**, **w**, and **z**). What happens if **u**, **v**, and **w** are coplanar?

1.12   Find an expression for a unit vector that lies in the intersection of the plane of **u** and **v** with the plane of **x** and **y**.

Reference

Hoffmann B 1975 *About Vectors*. Dover.

2

Tensors

Force and velocity can be described as vectors but other elements of continuum mechanics are described by tensors. There are many ways to define tensors and the subject is a rich one. Here we take a pragmatic, operational point of view: a tensor is defined in terms of its action on a vector. The quantities represented as tensors in continuum mechanics are physical entities. Consequently, as for vectors in Chapter 1, tensors are introduced in a coordinate-free form.

A *tensor* is a linear, homogeneous, vector-valued vector function. "Vector-valued vector function" means that a tensor acts on a vector and produces a vector as a result of the operation depicted schematically in Figure 2.1. Hence, the action of a tensor $\mathbf{F}$ on a vector $\mathbf{u}$ results in another vector $\mathbf{v}$:

$$\mathbf{v} = \mathbf{F}(\mathbf{u}) \tag{2.1}$$

"Homogeneous" (of degree 1) means that the function $\mathbf{F}$ has the property

$$\mathbf{F}(\alpha\mathbf{u}) = \alpha\mathbf{F}(\mathbf{u}) = \alpha\mathbf{v} \tag{2.2}$$

where α is a scalar. (Note: A scalar function $f(x, y)$ is said to be homogeneous of degree n if $f(\alpha x, \alpha y) = \alpha^n f(x, y)$. A function $f(x, y)$ is linear if

$$f(x, y) = \alpha x + \beta y + c$$

Hence, $f(x, y) = \sqrt{x^2 + y^2}$ is homogeneous of degree 1 but not linear. Similarly, $f(x, y) = a(x + y) + c$ is linear but not homogeneous.) The function $\mathbf{F}$ is "linear" if

$$\mathbf{F}(\mathbf{u}_1 + \mathbf{u}_2) = \mathbf{F}(\mathbf{u}_1) + \mathbf{F}(\mathbf{u}_2) = \mathbf{v}_1 + \mathbf{v}_2 \tag{2.3}$$

where $\mathbf{v}_1 = \mathbf{F}(\mathbf{u}_1)$ and $\mathbf{v}_2 = \mathbf{F}(\mathbf{u}_2)$. Combining the properties (2.1), (2.2), and (2.3) yields

$$\mathbf{F}(\alpha\mathbf{u}_1 + \beta\mathbf{u}_2) = \alpha\mathbf{F}(\mathbf{u}_1) + \beta\mathbf{F}(\mathbf{u}_2) = \alpha\mathbf{v}_1 + \beta\mathbf{v}_2 \tag{2.4}$$

Fundamentals of Continuum Mechanics, First Edition. John W. Rudnicki.
© 2015 John Wiley & Sons, Ltd. Published 2015 by John Wiley & Sons, Ltd.

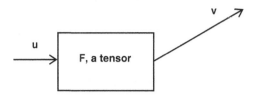

Figure 2.1 Schematic illustration of the action of a tensor on a vector. The tensor acts on the vector **u** and outputs the vector **v**.

where α and β are scalars. To determine if a "black box," a function **F**, is a tensor, we input vectors. If the results obey (2.4) then **F** must be a tensor. The definition of a tensor here is purely operational. It must pass the "duck test": If it has feathers like a duck, quacks like a duck, and walks like a duck, then we agree that it is a duck without the need to go further into what constitutes a duck.

The properties just discussed suggest that the action of a tensor on a vector can be represented as

$$\mathbf{v} = \mathbf{F} \cdot \mathbf{u} \tag{2.5}$$

The operation denoted by the dot is defined by the properties (2.2) and (2.3), or (2.4). The notation is meant to emphasize the connection with the analogous relation (1.2) for the dot product between two vectors.

Generally, the output vector **v** will have a different magnitude and direction from the input vector **u**. In the special case where the output vector is identical to the input vector, then, for obvious reasons, the tensor is called the *identity* tensor and denoted **I**. Hence, the identity tensor is defined by

$$\mathbf{u} = \mathbf{I} \cdot \mathbf{u} \tag{2.6}$$

for all vectors **u**.

Since both sides of (2.5) are vectors, we can form the scalar product with another vector, say **w**,

$$\mathbf{w} \cdot \mathbf{v} = \mathbf{w} \cdot (\mathbf{F} \cdot \mathbf{u})$$

and the result must be a scalar. Because scalar multiplication of two vectors is commutative, the order of the vectors on the left side can be reversed. On the right side, it would be necessary to write $(\mathbf{F} \cdot \mathbf{u}) \cdot \mathbf{w}$. The parentheses indicate that the operation $\mathbf{F} \cdot \mathbf{u}$ must be done first. If the parentheses were absent and the product $\mathbf{u} \cdot \mathbf{w}$ done first, the result would be a scalar. Because the dot is reserved for operations between vectors and tensors, the scalar product of a scalar with a vector (or a tensor) is not an operation that is defined. If, however, we define multiplication from the left by **w** as

$$\mathbf{w} \cdot (\mathbf{F} \cdot \mathbf{u}) = (\mathbf{w} \cdot \mathbf{F}) \cdot \mathbf{u}$$

then the result can be written without parentheses as

$$\mathbf{w} \cdot \mathbf{v} = \mathbf{w} \cdot \mathbf{F} \cdot \mathbf{u} \qquad (2.7)$$

Thus, writing the result as on the right side of (2.7) makes the meaning clear even if the parentheses are omitted.

In contrast to the dot product of two vectors, the dot product of a tensor and a vector is not commutative. Reversing the order defines the *transpose* of the tensor $\mathbf{F}$, i.e.,

$$\mathbf{F} \cdot \mathbf{u} = \mathbf{u} \cdot \mathbf{F}^T \qquad (2.8)$$

Thus, it follows that

$$\mathbf{v} \cdot \mathbf{F} \cdot \mathbf{u} = \mathbf{u} \cdot \mathbf{F}^T \cdot \mathbf{v}$$

where parentheses are not needed, as just explained. If $\mathbf{F} = \mathbf{F}^T$, then the tensor $\mathbf{F}$ is said to be *symmetric*; if $\mathbf{F} = -\mathbf{F}^T$, then $\mathbf{F}$ is *antisymmetric* or *skew-symmetric*. Every tensor can be separated into the sum of a symmetric and a skew-symmetric tensor by adding and subtracting half of its transpose

$$\mathbf{F} = \frac{1}{2}\left(\mathbf{F} + \mathbf{F}^T\right) + \frac{1}{2}\left(\mathbf{F} - \mathbf{F}^T\right) \qquad (2.9)$$

2.1 Inverse

Is it possible to operate our tensor black box in reverse? In terms of Figure 2.1, if we insert $\mathbf{v}$ in the right side, will we get $\mathbf{u}$ out the left? The answer is "not always," although for the particular tensors we are concerned with it will be possible in most cases. Later we will determine the conditions for which the operation depicted in Figure 2.1 is reversible. If it is, then the operation defines the inverse of $\mathbf{F}$

$$\mathbf{u} = \mathbf{F}^{-1} \cdot \mathbf{v} \qquad (2.10)$$

Substituting for $\mathbf{v}$ from (2.5) and using (2.6) gives

$$\left\{\mathbf{F}^{-1} \cdot \mathbf{F} - \mathbf{I}\right\} \cdot \mathbf{u} = 0$$

Because this relation applies for *any* vector $\mathbf{u}$, the expression in the braces must vanish, giving

$$\mathbf{F}^{-1} \cdot \mathbf{F} = \mathbf{I} \qquad (2.11)$$

2.2   Orthogonal Tensor

If the output vector $\mathbf{v}$ has the same magnitude as the input vector $\mathbf{u}$, but a different direction, then the tensor operation results in a rotation

$$\mathbf{v} = \mathbf{A} \cdot \mathbf{u} \tag{2.12}$$

and the tensor is called *orthogonal*. Because $\mathbf{u}$ and $\mathbf{v}$ have the same magnitude

$$v^2 = \mathbf{v} \cdot \mathbf{v} = \mathbf{u} \cdot \mathbf{u} = u^2$$

Using (2.8) to rewrite the left scalar product and (2.6) to rewrite the right gives

$$\mathbf{u} \cdot \mathbf{A}^T \cdot \mathbf{A} \cdot \mathbf{u} = \mathbf{u} \cdot \mathbf{I} \cdot \mathbf{u} \tag{2.13}$$

where again no parentheses are necessary. Because (2.13) applies for *any* vector $\mathbf{u}$, we can conclude that

$$\mathbf{A}^T \cdot \mathbf{A} = \mathbf{I} \tag{2.14}$$

Comparing to (2.11) reveals that the transpose of an orthogonal tensor is equal to its inverse. Physically, the rotation of a vector to another direction can always be reversed, so we expect the inverse of an orthogonal tensor to exist.

2.3   Principal Values

Is it possible to find an input vector $\mathbf{u}$ such that the output vector $\mathbf{v}$ has the same direction, but a different magnitude? Intuitively, we expect that this is only possible for certain input vectors, if any. If the vector $\mathbf{v}$ is in the same direction as $\mathbf{u}$, then $\mathbf{v} = \lambda\mathbf{u}$, where λ is a scalar. Substituting in (2.5) yields

$$\mathbf{F} \cdot \mathbf{u} = \lambda\mathbf{u} \tag{2.15}$$

or, after using (2.6),

$$(\mathbf{F} - \lambda\mathbf{I}) \cdot \mathbf{u} = \mathbf{0} \tag{2.16}$$

If the inverse of $\mathbf{F} - \lambda\mathbf{I}$ exists then the only possible solution is $\mathbf{u} = \mathbf{0}$. Consequently there will be special values of λ and $\mathbf{u}$ that satisfy this equation only when the inverse does not exist. A value of λ that does so is a *principal value* (*eigenvalue*) of the tensor $\mathbf{F}$ and the corresponding direction given by $\mathbf{u}$ is the *principal direction* (*eigenvector*). It is clear from (2.16) that if $\mathbf{u}$ is a solution, then so is $\alpha\mathbf{u}$ where α is any scalar. Hence, only the direction of the eigenvector is determined. Thus it is always possible to normalize the eigenvector to unit magnitude, $\mu = \mathbf{u}/u$.

Later we will learn how to determine the principal values and directions and their physical significance. But, because all of the tensors we will deal with are real and many of them are symmetric, we can prove that the eigenvalues and eigenvectors must have certain properties without having to determine them explicitly.

First we will prove that a real symmetric tensor has real principal values. Let $\mathbf{F}$ be a real symmetric order tensor with a principal value λ and corresponding principal direction μ satisfying

$$\mathbf{F} \cdot \mu = \lambda\mu \tag{2.17}$$

Taking the complex conjugate of both sides gives

$$\bar{\mathbf{F}} \cdot \bar{\mu} = \bar{\lambda}\bar{\mu} \tag{2.18}$$

(Taking the conjugate of a complex expression means changing the sign of $i = \sqrt{-1}$ wherever it appears.) Multiplying (2.17) by $\bar{\mu}$ yields

$$\bar{\mu} \cdot \mathbf{F} \cdot \mu = \lambda\bar{\mu} \cdot \mu \tag{2.19}$$

and (2.18) by μ yields

$$\mu \cdot \bar{\mathbf{F}} \cdot \bar{\mu} = \bar{\lambda}\bar{\mu} \cdot \mu \tag{2.20}$$

Because $\mathbf{F}$ is real and symmetric, $\mathbf{F} = \bar{\mathbf{F}}^T$, and the left sides of (2.19) and (2.20) are the same. Subtracting gives

$$0 = (\lambda - \bar{\lambda})\mu \cdot \bar{\mu}$$

Since $\bar{\mu} \cdot \mu \neq 0$, $\lambda = \bar{\lambda}$ and hence the principal values are real.

Now we prove that the eigenvectors corresponding to distinct eigenvalues are orthogonal. For principal value λ_I and corresponding principal direction μ_I

$$\mathbf{F} \cdot \mu_I = \lambda_I\,\mu_I \tag{2.21}$$

and similarly for λ_{II} and μ_{II}

$$\mathbf{F} \cdot \mu_{II} = \lambda_{II}\,\mu_{II} \tag{2.22}$$

Forming the scalar product of (2.21) with μ_{II} and (2.22) with μ_I yields

$$\mu_{II} \cdot \mathbf{F} \cdot \mu_I = \lambda_I\,\mu_I \cdot \mu_{II} \tag{2.23}$$

$$\mu_I \cdot \mathbf{F} \cdot \mu_{II} = \lambda_{II}\,\mu_{II} \cdot \mu_I \tag{2.24}$$

Because $\mathbf{F} = \mathbf{F}^T$ the left sides of (2.23) and (2.24) are equal and subtraction yields

$$(\lambda_I - \lambda_{II})\mu_I \cdot \mu_{II} = 0$$

Because the principal values are assumed to be distinct, $\lambda_I \neq \lambda_{II}$, and consequently $\mu_I \cdot \mu_{II} = 0$. If two of the principal values are equal, say $\lambda_I = \lambda_{II}$, but distinct from the third, then any vectors in the plane perpendicular to the principal direction of the third principal value can serve as principal directions. Therefore, it is always possible to find at least one set of orthogonal eigenvectors.

2.4   Nth-Order Tensors

Lastly, we note that the tensors we have introduced here are *second-order tensors* because they input a vector and output a vector. We can, however, define nth-order tensors $\mathbf{F}^{(n)}$ by the following recursive relation:

$$\mathbf{F}^{(n)} \cdot \mathbf{u} = \mathbf{F}^{(n-1)} \tag{2.25}$$

If $\mathbf{F}^{(0)}$ is defined as a scalar then (2.25) shows that a vector can be considered as a tensor of order 1. Later we will have occasion to deal with third- and fourth-order tensors.

2.5   Examples

2.5.1

Show that the product $\mathbf{H} = \mathbf{F} \cdot \mathbf{G}$ is a tensor where $\mathbf{F}$ and $\mathbf{G}$ are tensors.

Because $\mathbf{F}$ is a tensor it satisfies (2.5) and because $\mathbf{G}$ is a tensor, $\mathbf{b} = \mathbf{G} \cdot \mathbf{a}$ for vectors $\mathbf{a}$ and $\mathbf{b}$. Letting $\mathbf{u} = \mathbf{b}$ in (2.5) yields

$$\mathbf{v} = \mathbf{F} \cdot \mathbf{G} \cdot \mathbf{a} = \mathbf{H} \cdot \mathbf{a}$$

Letting $\mathbf{a} = \alpha_1 \mathbf{a}_1 + \alpha_2 \mathbf{a}_2$ and using the properties implied by the dot verifies that $\mathbf{H}$ satisfies (2.4) and, thus, is a tensor.

2.5.2

If $\mathbf{F}$ and $\mathbf{G}$ are tensors, use the result from Example 2.5.1 and the definition of the transpose (2.8) to show that

$$(\mathbf{F} \cdot \mathbf{G})^T = \mathbf{G}^T \cdot \mathbf{F}^T$$

By the definition of the transpose (2.8)

$$\mathbf{H} \cdot \mathbf{u} = \mathbf{u} \cdot \mathbf{H}^T$$

Letting $\mathbf{H} = \mathbf{F} \cdot \mathbf{G}$ yields

$$\mathbf{F} \cdot \mathbf{G} \cdot \mathbf{u} = \mathbf{u} \cdot (\mathbf{F} \cdot \mathbf{G})^T$$

using the definition of the transpose on $\mathbf{G} \cdot \mathbf{u} = \mathbf{u} \cdot \mathbf{G}^T$. But $\mathbf{v} = \mathbf{G} \cdot \mathbf{u} = \mathbf{u} \cdot \mathbf{G}^T$ is itself a vector. Therefore,

$$\mathbf{F} \cdot \mathbf{v} = \mathbf{v} \cdot \mathbf{F}^T$$

Substituting $\mathbf{v} = \mathbf{u} \cdot \mathbf{G}^T$ on the right side establishes the result.

Exercises

2.1 In the additive decomposition of a tensor

$$\mathbf{F} = \frac{1}{2} \left(\mathbf{F} + \mathbf{F}^T \right) + \frac{1}{2} \left(\mathbf{F} - \mathbf{F}^T \right)$$

verify that the first term is symmetric and the second term is antisymmetric.

2.2 Show the result analogous to (2.11) if the inverse is on the right:

$$\mathbf{F} \cdot \mathbf{F}^{-1} = \mathbf{I}$$

2.3 If $\mathbf{F}$ and $\mathbf{G}$ are tensors, use the result from Example 2.5.1 and the definition of the inverse (2.10) to show that

$$(\mathbf{F} \cdot \mathbf{G})^{-1} = \mathbf{G}^{-1} \cdot \mathbf{F}^{-1}$$

2.4 If $\mathbf{F}$ is a tensor, show that

$$\left(\mathbf{F}^{-1} \right)^T = \left(\mathbf{F}^T \right)^{-1}$$

2.5 If $\mathbf{R}$ and $\mathbf{S}$ are orthogonal tensors show that the product $\mathbf{R} \cdot \mathbf{S}$ is also orthogonal.

2.6 Show that the angle between vectors $\mathbf{u}$ and $\mathbf{v}$ is identical to the angle between the vectors that result from application of the orthogonal tensor $\mathbf{A}$ separately to $\mathbf{u}$ and $\mathbf{v}$.

3

Cartesian Coordinates

Chapter 1 and Chapter 2 introduced vectors and tensors in coordinate-free notation: that is, without referring to any particular coordinate system. Philosophically, this is attractive because it emphasizes the independence of the physical entities described by vectors and tensors from their description in a particular system. Defining the rules for this description soon becomes cumbersome, however, and it is convenient to discuss vectors and tensors in terms of their components in a coordinate system. Moreover, when considering a particular problem or implementing the formulation on a computer, it is necessary to adopt a coordinate system.

Given that a coordinate system is necessary, we might take the approach that we should express our results in a form that is appropriate for completely arbitrary coordinate systems. That is, we could make no assumptions that the axes of the system are orthogonal or scaled in the same way and so on. This is often useful and can lead to a deeper understanding of vectors and tensors. (Section 4.5 gives a brief introduction to this approach.) Nevertheless, it requires the introduction of many details that, at least at this stage, will be distracting.

Consequently, we will consider almost exclusively rectangular Cartesian coordinate systems. We will, however, continue to use and emphasize a coordinate-free notation. Fortunately, results that can be expressed in a coordinate-free notation, if interpreted properly, can be translated into components in any arbitrary coordinate system.

3.1 Base Vectors

A rectangular Cartesian coordinate system with origin O is shown in Figure 3.1. The axes are orthogonal and are labeled x, y, and z, or x_1, x_2, and x_3. A convenient way to specify the coordinate system is to introduce vectors that are tangent to the coordinate directions. More generally, a set of vectors is a *basis* for the space (here three-dimensional) if every vector in the space can be expressed as a unique linear combination of the base vectors. For rectangular Cartesian systems, it is convenient to use unit vectors as base vectors

$$|\mathbf{e}_1| = \mathbf{e}_1 \cdot \mathbf{e}_1 = 1, \quad |\mathbf{e}_2| = |\mathbf{e}_3| = 1 \tag{3.1}$$

Fundamentals of Continuum Mechanics, First Edition. John W. Rudnicki.
© 2015 John Wiley & Sons, Ltd. Published 2015 by John Wiley & Sons, Ltd.

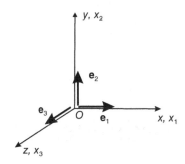

Figure 3.1   Rectangular Cartesian coordinate system specified by unit, orthogonal base vectors.

that are orthogonal:

$$\mathbf{e}_1 \cdot \mathbf{e}_2 = 0, \quad \mathbf{e}_1 \cdot \mathbf{e}_3 = 0, \quad \mathbf{e}_2 \cdot \mathbf{e}_3 = 0 \tag{3.2}$$

The six equations, (3.1) and (3.2), and the additional three that result from reversing the order of the scalar product in (3.2) can be written compactly as

$$\mathbf{e}_i \cdot \mathbf{e}_j = \delta_{ij} = \begin{cases} 1 & \text{if} \quad i = j \\ 0 & \text{if} \quad i \neq j \end{cases} \tag{3.3}$$

where the indices (i, j) stand for $(1, 2, 3)$ and δ_{ij} is the *Kronecker delta*. Therefore, (3.3) represents nine equations. Note that one i and one j appear on each side of the equation and that each index is free to take on the value 1, 2, or 3. Consequently, i and j in (3.3) are *free indices*.

A scalar component of the vector $\mathbf{u}$ is given by its projection on a coordinate direction:

$$u_i = \mathbf{e}_i \cdot \mathbf{u} \tag{3.4}$$

Equation (3.4) stands for three equations, one for $i = 1, 2, 3$. Because $\mathbf{e}_i$ is a unit vector, (1.2) indicates that the right side of (3.4) is the magnitude of $\mathbf{u}$ multiplied by the cosine of the angle between $\mathbf{u}$ and $\mathbf{e}_i$. We can now represent the vector $\mathbf{u}$ as the sum of the products of the scalar components with the unit base vectors:

$$\mathbf{u} = u_1 \mathbf{e}_1 + u_2 \mathbf{e}_2 + u_3 \mathbf{e}_3 \tag{3.5}$$

Each term, e.g., $u_1 \mathbf{e}_1$, is a vector component of $\mathbf{u}$. The left side of (3.5) is a coordinate-free representation: that is, it makes no reference to a particular coordinate system that we are using to represent the vector. The right side is the component form; the presence of the base vectors $\mathbf{e}_1$, $\mathbf{e}_2$, and $\mathbf{e}_3$ denotes explicitly that u_1, u_2, and u_3 are the components with respect to the coordinate system with these particular base vectors. For a different coordinate system, with different base vectors, the right side would be different but would still represent the same vector, indicated by the coordinate-free form on the left side.

3.2   Summation Convention

Equation (3.5) can be expressed more concisely by using the summation sign:

$$\mathbf{u} = \sum_{k=1}^{3} u_k \mathbf{e}_k = u_k \mathbf{e}_k \tag{3.6}$$

where k takes on the explicit values 1, 2, and 3. Consequently, it is called a *summation* or *dummy* index because it is simply a placeholder: changing k to m does not alter the meaning of the equation. (In contrast, the free index i on the right side of (3.4) cannot be changed to m without making the same change on the other side of the equation.) Note that in (3.6) k appears twice on the right side but not on the left. Because the form (3.6) occurs so frequently, we will adopt the *summation convention*: The summation symbol is dropped and summation is implied whenever an index is repeated in an additive term (a term separated by a plus or minus sign) on one side of the equation. This is a compact and powerful notation but it requires adherence to certain rules. Regardless of the physical meaning of the equation, the following rules apply:

- A subscript should never appear more than exactly twice (in each additive term) on one side of an equation.
- If a subscript appears once on one side of an equation it must appear exactly once (in each additive term) on the other side

For example, both of the following two equations are incorrect because the index j appears once on the right side but not at all on the left:

$$w_i = u_i + v_j$$

$$w_i = u_k v_j s_k t_i$$

The following equation is incorrect because the index k appears three times in an additive term:

$$w_{ij} = A_{ik} B_{jk} u_k \tag{3.7}$$

In contrast, the equation

$$a = u_k v_k + r_k s_k + p_k q_k$$

is correct. Even though k appears six times on the right side, it only appears twice in each additive term.

The default interpretation is that a repeated index implies summation. Consequently, in an expression that contains a repeated index that is not meant to be summed, it must be denoted explicitly; for example, "No sum on k" if k is the repeated index. The summation convention applies only to the repeat of two indices. Consequently, if three indices occur (in an additive term), such as in (3.7), either one of the indices must be changed or, if they are to be summed, the summation must then be indicated explicitly.

Although these rules are simple, it is important to adhere to them as indices will multiply faster than rabbits in succeeding chapters. A familiarity with the application of these rules provides guidance not only to the manipulation of expressions but also to their meaning.

To reproduce (3.4), the component of the vector $\mathbf{u}$ with respect to the ith coordinate direction, we form the scalar product $\mathbf{e}_i \cdot \mathbf{u}$ and then express $\mathbf{u}$ in its component form:

$$\mathbf{e}_i \cdot \mathbf{u} = \mathbf{e}_i \cdot (u_j \mathbf{e}_j)$$

Note that it would be incorrect to write $u_i \mathbf{e}_i$ on the right side since the index i would then appear three times. The scalar product is an operation between vectors and, thus, applies to the two base vectors. The result is given by (3.3). Recalling that the repeated j implies summation and writing the terms explicitly gives

$$\mathbf{e}_i \cdot \mathbf{u} = u_j \delta_{ij} = \sum_{j=1}^{3} \delta_{ij} u_j = \delta_{i1} u_1 + \delta_{i2} u_2 + \delta_{i3} u_3 = u_i$$

We can now use the scalar product, base vectors, and index notation to convert some of the coordinate-free relations in Chapter 1 to component form. For example, the sum of two vectors is given by (1.1) in the coordinate-free notation. Forming the scalar product of both sides with the base vectors $\mathbf{e}_i$ yields the component form

$$w_i = u_i + v_i$$

Thus, in a rectangular Cartesian coordinate system, the component of the sum of two vectors is the sum of the corresponding components of the two vectors.

As a final example, we derive the expression for the scalar product $\mathbf{u} \cdot \mathbf{v}$ in terms of the components of the vectors. Substituting the component representations, noting that the scalar product is an operation between vectors, and using (3.3) yields

$$\mathbf{u} \cdot \mathbf{v} = u_i v_j \delta_{ij} = \sum_{i=1}^{3} \sum_{j=1}^{3} u_i v_j \delta_{ij} = \sum_{j=1}^{3} u_j v_j = u_j v_j$$

3.3   Tensor Components

The definition of a tensor embodied by the properties (2.1), (2.2), and (2.3) or (2.4) suggests that the action of tensor on a vector can be represented in coordinate-free notation by (2.5). The Cartesian component or index representation follows from the procedure for identifying the Cartesian components of vectors, i.e.,

$$v_k = \mathbf{e}_k \cdot \mathbf{v} = \mathbf{e}_k \cdot \{\mathbf{F} \cdot u_l \mathbf{e}_l\}$$
$$= (\mathbf{e}_k \cdot \mathbf{F} \cdot \mathbf{e}_l) u_l$$

The second line above can be represented in the component form

$$v_k = F_{kl} u_l \tag{3.8}$$

or in the matrix form

$$\begin{bmatrix} v_1 \\ v_2 \\ v_3 \end{bmatrix} = \begin{bmatrix} F_{11} & F_{12} & F_{13} \\ F_{21} & F_{22} & F_{23} \\ F_{31} & F_{32} & F_{33} \end{bmatrix} \begin{bmatrix} u_1 \\ u_2 \\ u_3 \end{bmatrix} \tag{3.9}$$

where

$$F_{kl} = \mathbf{e}_k \cdot \mathbf{F} \cdot \mathbf{e}_l \tag{3.10}$$

are the Cartesian components of the tensor $\mathbf{F}$ (with respect to the base vectors $\mathbf{e}_l$). Although (3.9) is a convenient representation for computations, a disadvantage is that it does not contain information about the appropriate base vectors.

The expression (3.10) for the Cartesian components of a tensor leads naturally to the representation of tensors as

$$\mathbf{F} = F_{kl}\mathbf{e}_k\mathbf{e}_l \tag{3.11}$$

Substituting this form into (3.10) gives an identity simply by using the rules that have already been established for vectors and the properties of the Kronecker delta (3.3). The matrix equation (3.9) can also be written as

$$[v_1 \ \ v_2 \ \ v_3] = [u_1 \ \ u_2 \ \ u_3] \begin{bmatrix} F_{11} & F_{21} & F_{31} \\ F_{12} & F_{22} & F_{32} \\ F_{13} & F_{23} & F_{33} \end{bmatrix}$$

or more compactly as $[v]^T = [u]^T[F]^T$. Note, however, in the index notation (3.8), or in the dyadic notation to be discussed next, that there is no need for the transpose of a vector corresponding to change from a column vector to a row vector.

3.4 Dyads

Equation (3.11) represents a tensor as a *dyadic*, a polynomial of dyads. A *dyad* is two vectors placed next to each other, e.g., $\mathbf{ab}$, $\mathbf{e}_1\mathbf{e}_2$, $\mathbf{ij}$ (although the notation $\mathbf{a} \otimes \mathbf{b}$ is often used). The meaning of a dyad is defined operationally by its action on a vector:

$$(\mathbf{ab}) \cdot \mathbf{v} = \mathbf{a}(\mathbf{b} \cdot \mathbf{v}) \tag{3.12}$$

Because a dyad operates on a vector and outputs a vector by means of a rule that is linear and homogeneous, it can be used to represent a tensor. Although this notation may appear strange, it can be given a simple interpretation in matrix notation:

$$\mathbf{ab} = \begin{bmatrix} a_1 \\ a_2 \\ a_3 \end{bmatrix} [b_1 \ \ b_2 \ \ b_3] = \begin{bmatrix} a_1b_1 & a_1b_2 & a_1b_3 \\ a_2b_1 & a_2b_2 & a_2b_3 \\ a_3b_1 & a_3b_2 & a_3b_3 \end{bmatrix}$$

Hence, the operation indicated by (3.12) written in matrix form is

$$\begin{bmatrix} a_1b_1 & a_1b_2 & a_1b_3 \\ a_2b_1 & a_2b_2 & a_2b_3 \\ a_3b_1 & a_3b_2 & a_3b_3 \end{bmatrix} \begin{bmatrix} v_1 \\ v_2 \\ v_3 \end{bmatrix} = \begin{bmatrix} a_1 \\ a_2 \\ a_3 \end{bmatrix} [b_1v_1 + b_2v_2 + b_3v_3]$$

Equation (3.12) implies that multiplication by a dyad is not commutative, i.e.,

$$\mathbf{v} \cdot (\mathbf{ab}) = \mathbf{b}(\mathbf{v} \cdot \mathbf{a}) \tag{3.13}$$

The transpose of a dyad is defined by reversing the order of the vectors that make up the dyad. Thus, the transpose of the dyad $\mathbf{ab}$ is $\mathbf{ba}$. In matrix form (3.13) can be written as

$$[v_1 \quad v_2 \quad v_3] \begin{bmatrix} a_1b_1 & a_1b_2 & a_1b_3 \\ a_2b_1 & a_2b_2 & a_2b_3 \\ a_3b_1 & a_3b_2 & a_3b_3 \end{bmatrix} = \begin{bmatrix} a_1b_1 & a_1b_2 & a_1b_3 \\ a_2b_1 & a_2b_2 & a_2b_3 \\ a_3b_1 & a_3b_2 & a_3b_3 \end{bmatrix}^T \begin{bmatrix} v_1 \\ v_2 \\ v_3 \end{bmatrix}$$

Equation (3.11) and

$$\mathbf{\Phi} = \mathbf{a}_1\mathbf{b}_1 + \mathbf{a}_2\mathbf{b}_2 + \mathbf{a}_3\mathbf{b}_3 \tag{3.14}$$

are examples of dyadics. The transpose of a dyadic reverses each pair of vectors, e.g.,

$$\mathbf{\Phi}^T = \mathbf{b}_1\mathbf{a}_1 + \mathbf{b}_2\mathbf{a}_2 + \mathbf{b}_3\mathbf{a}_3$$

Application of this rule to (3.11) is consistent with the definition of the transpose of a tensor given by (2.8):

$$\mathbf{F}^T = F_{ij}\mathbf{e}_j\mathbf{e}_i = F_{qp}\mathbf{e}_p\mathbf{e}_q \tag{3.15}$$

In the preceding equation, the second equality follows by relabeling i as q and j as p (permissible because these are summation or dummy indices). As a consequence $(\mathbf{F}^T)_{ij} = F_{ji}$ and the transpose in (3.15) can be formed by reversing the order of the base vectors or the indices, but not both.

Multiplication of the dyadic (3.14) by a vector is given by

$$\mathbf{v} \cdot \mathbf{\Phi} = (\mathbf{v} \cdot \mathbf{a}_1)\mathbf{b}_1 + (\mathbf{v} \cdot \mathbf{a}_2)\mathbf{b}_2 + (\mathbf{v} \cdot \mathbf{a}_3)\mathbf{b}_3$$

Multiplication is distributive:

$$(\mathbf{a} + \mathbf{b})(\mathbf{c} + \mathbf{d}) = \mathbf{ac} + \mathbf{bc} + \mathbf{ad} + \mathbf{bd}$$

$$\mathbf{\Phi} = \mathbf{ab} = (a_k\mathbf{e}_k)(b_l\mathbf{e}_l) = a_kb_l\mathbf{e}_k\mathbf{e}_l$$

As for a tensor, a dyadic is symmetric if it is equal to its transpose and (3.15) shows that the components of the tensor satisfy

$$F_{ij} = F_{ji}$$

Similarly, a dyadic or tensor is antisymmetric if it is equal to the negative of its transpose. Thus, the components satisfy

$$F_{ij} = -F_{ji}$$

As a result the diagonal components are all zero

$$F_{11} = F_{22} = F_{33} = 0$$

and the off-diagonal components are the negative of each other,

$$F_{21} = -F_{12}, \quad F_{13} = -F_{31}, \quad F_{23} = -F_{32}$$

As noted earlier (2.9), any second-order tensor can be written as the sum of a symmetric and antisymmetric part.

The identity tensor $\mathbf{I}$ in (2.6) was defined as that tensor whose product with a vector gives the identical vector. This implies that $\mathbf{I}$ has the following dyadic representation in terms of orthonormal base vectors:

$$\mathbf{I} = \delta_{mn}\mathbf{e}_m\mathbf{e}_n = \mathbf{e}_1\mathbf{e}_1 + \mathbf{e}_2\mathbf{e}_2 + \mathbf{e}_3\mathbf{e}_3$$

3.5 Tensor and Scalar Products

As shown in Example 2.5.1, the tensor product of two tensors $\mathbf{F}$ and $\mathbf{G}$ is itself a tensor. The Cartesian component or index form of the product tensor $\mathbf{H} = \mathbf{F} \cdot \mathbf{G}$ is defined naturally using the dyadic representation and operations between the base vectors:

$$\mathbf{F} \cdot \mathbf{G} = (F_{ij}\mathbf{e}_i\mathbf{e}_j) \cdot (G_{kl}\mathbf{e}_k\mathbf{e}_l) \tag{3.16}$$

$$= F_{ij}G_{kl}\mathbf{e}_i(\mathbf{e}_j \cdot \mathbf{e}_k)\mathbf{e}_l \tag{3.17}$$

$$= F_{ik}G_{kl}\mathbf{e}_i\mathbf{e}_l \tag{3.18}$$

Using (3.10) gives the scalar components of $\mathbf{H}$ as

$$H_{il} = F_{ik}G_{kl} \tag{3.19}$$

Note that (3.16) to (3.18) are single tensor or dyadic equations; (3.19) represents nine scalar equations for the Cartesian components of $\mathbf{H}$.

The components of the product tensor can be computed in the usual way by matrix multiplication of the components of $\mathbf{F}$ and $\mathbf{G}$:

$$F_{ik}G_{kl} = \begin{bmatrix} F_{11} & F_{12} & F_{13} \\ F_{21} & F_{22} & F_{23} \\ F_{31} & F_{32} & F_{33} \end{bmatrix} \begin{bmatrix} G_{11} & G_{12} & G_{13} \\ G_{21} & G_{22} & G_{23} \\ G_{31} & G_{32} & G_{33} \end{bmatrix}$$

As with matrix multiplication, the tensor product is not commutative. In fact, the rule that the transpose of a matrix product is the product of the transposes of the individual matrices is easily verified by the rules for computing with the components of the dyad:

$$\mathbf{F} \cdot \mathbf{G} = \{\mathbf{G}^T \cdot \mathbf{F}^T\}^T \tag{3.20}$$

Example 2.5.2 proves this in coordinate-free form.

Index notation for Cartesian coordinate systems can be used to prove a relation between two coordinate-free representations. Because the resulting coordinate-free forms are valid in any coordinate system, they can, with appropriate interpretation, be expressed in terms of non-Cartesian components.

The scalar product between two tensors can be computed in two ways depending on the order in which the dot products between the base vectors are taken. Notation varies on this product but here we follow Malvern (1988) and use the horizontal arrangement of the dots to indicate that the dot product is taken between the two closest base vectors (the two inside) and then the two furthest (the two outside):

$$\mathbf{F} \cdot\cdot \mathbf{G} = (F_{ij}\mathbf{e}_i\mathbf{e}_j) \cdot\cdot (G_{kl}\mathbf{e}_k\mathbf{e}_l)$$
$$= F_{ij}G_{kl}(\mathbf{e}_i\cdot\mathbf{e}_l)(\mathbf{e}_j\cdot\mathbf{e}_k)$$
$$= F_{lk}G_{kl}$$

Malvern (1988) uses vertical dots to indicate that the first base vectors of each dyad are dotted and the second base vectors of each are dotted:

$$\mathbf{F} : \mathbf{G} = (F_{ij}\mathbf{e}_i\mathbf{e}_j) : (G_{kl}\mathbf{e}_k\mathbf{e}_l)$$
$$= F_{ij}G_{kl}(\mathbf{e}_i\cdot\mathbf{e}_k)(\mathbf{e}_j\cdot\mathbf{e}_l)$$
$$= F_{lk}G_{lk} \tag{3.21}$$

But the same result is obtained by using the transpose of one of the tensors in (3.21). If either of the tensors is symmetric then the two scalar products are identical.

The trace of a tensor $\mathbf{F}$ is a scalar obtained by forming the scalar product of $\mathbf{F}$ with the identity tensor:

$$\text{tr}(\mathbf{F}) = \mathbf{F} \cdot\cdot \mathbf{I} \tag{3.22}$$

3.6   Examples

3.6.1

Example 2.5.2 is to prove (3.20) without recourse to component representation. To prove it using dyadic or component form:

$$
\begin{aligned}
\mathbf{F} \cdot \mathbf{G} &= F_{ij}\mathbf{e}_i\mathbf{e}_j \cdot G_{kl}\mathbf{e}_k\mathbf{e}_l \\
&= \mathbf{e}_i F_{ik} G_{kl} \mathbf{e}_l \\
&= \{\mathbf{e}_l F_{ik} G_{kl} \mathbf{e}_i\}^T \\
&= \{\mathbf{e}_l G_{kl}\mathbf{e}_k \cdot \mathbf{e}_j F_{ij}\mathbf{e}_i\}^T \\
&= \{G_{kl}\mathbf{e}_l\mathbf{e}_k \cdot F_{ij}\mathbf{e}_j\mathbf{e}_i\}^T \\
&= \{\mathbf{G}^T \cdot \mathbf{F}^T\}^T
\end{aligned}
$$

3.6.2

Use dyadic or index (Cartesian component) notation to show that $\mathbf{F} \cdot \cdot \mathbf{G}^T = \mathbf{F}^T \cdot \cdot \mathbf{G}$.
   Here

$$
\begin{aligned}
\mathbf{F} \cdot \cdot \mathbf{G}^T &= F_{ij} G_{ij} \\
&= F_{ji}^T G_{ij} = \mathbf{F}^T \cdot \cdot \mathbf{G}
\end{aligned}
$$

3.6.3

If $\mathbf{F}$ is a symmetric tensor and $\mathbf{G}$ is a skew-symmetric tensor, use dyadic or index (Cartesian component) notation to show that

$$
\mathbf{F} \cdot \cdot \mathbf{G} = 0
$$

   Here

$$
\begin{aligned}
\mathbf{F} \cdot \cdot \mathbf{G} &= F_{ij} G_{ji} \\
&= \frac{1}{2}\{F_{ij} G_{ji} - F_{ij} G_{ji}^T\} \\
&= \frac{1}{2}\{F_{ij} G_{ji} - F_{ij} G_{ij}\} \\
&= \frac{1}{2}\{F_{ij} G_{ji} - F_{ji}^T G_{ij}\} \\
&= \frac{1}{2}\{F_{ij} G_{ji} - F_{ji} G_{ij}\} = 0
\end{aligned}
$$

Exercises

3.1 Write out the following expressions completely (i.e., replace all indices by appropriate numbers):

(a) $v_i = H_{ji}u_j$

(b) $e = e_{kk}$

(c) $W = F_{ij}H_{ij}$

(d) $F_{ij} = aG_{ij}$

3.2 Verify the following identities by writing out all terms:

(a) $\delta_{mm} = 3$

(b) $\delta_{mn}\delta_{mn} = 3$

3.3 Write out Problems 3.1.a and 3.1.d in matrix form.

3.4 Use the dyadic or index (Cartesian component) form of **I** to prove that

(a) $\mathbf{v} \cdot \mathbf{I} = \mathbf{v}$

(b) $\mathbf{F} \cdot \mathbf{I} = \mathbf{F}$

3.5 Use index notation to show that if $\mathbf{u} \cdot \mathbf{v} = 0$ for any vector **v** then each component of **u** must be zero.

3.6 Use the dyadic or index (Cartesian component) forms of **F** and **I** in (3.22) to show that the trace is equal to the sum of the three diagonal components: $\mathrm{tr}(\mathbf{F}) = F_{nn}$.

3.7 Use dyadic or index (Cartesian component) form to show that $\mathbf{F} \cdot\cdot \mathbf{G} = \mathbf{G} \cdot\cdot \mathbf{F}$.

3.8 Use dyadic or index (Cartesian component) form to show that

$$\mathbf{F} \cdot\cdot \mathbf{G} = \mathbf{F} : \mathbf{G}^T = \mathbf{F}^T : \mathbf{G}$$

3.9 For tensors **F**, **G**, and **H** use dyadic or index (Cartesian component) notation to prove that

$$\mathbf{F} \cdot \mathbf{G} \cdot\cdot \mathbf{H} = \mathbf{F} \cdot\cdot \mathbf{G} \cdot \mathbf{H}$$
$$= \mathbf{H} \cdot \mathbf{F} \cdot\cdot \mathbf{G}$$
$$= \mathbf{G} \cdot \mathbf{H} \cdot\cdot \mathbf{F}$$

3.10 Use dyadic or index (Cartesian component) notation to show that the results in the preceding problem do not apply if the horizontal dots are replaced by vertical dots.

3.11 Use dyadic or index (Cartesian component) notation to show that the trace can also be used to represent the scalar product

$$\mathbf{F} \cdot\cdot \mathbf{G} = \mathrm{tr}(\mathbf{F} \cdot \mathbf{G})$$

Reference

Malvern LE 1988 *Introduction to the Mechanics of a Continuous Medium*. Prentice Hall.

4

Vector (Cross) Product

We introduced the coordinate-free form of the vector or cross product in Chapter 1. Here we will introduce the component form.

For two vectors $\mathbf{u}$ and $\mathbf{v}$, there are nine (3^2) possible products of their components. The scalar product is the sum of three. The remaining six can be combined in pairs to form a vector:

$$\mathbf{w} = \mathbf{u} \times \mathbf{v} = (u_i \mathbf{e}_i) \times (v_j \mathbf{e}_j) = u_i v_j (\mathbf{e}_i \times \mathbf{e}_j) \tag{4.1}$$

To interpret (4.1), we first consider the cross products of the base vectors. The vector

$$\mathbf{e}_3 = \mathbf{e}_1 \times \mathbf{e}_2$$

is perpendicular to the plane containing $\mathbf{e}_1$ and $\mathbf{e}_2$ with the sense given by the right hand rule. Consequently, reversing the order of the two vectors in the product must change the sign:

$$\mathbf{e}_1 \times \mathbf{e}_2 = -\mathbf{e}_2 \times \mathbf{e}_1$$

Similarly,

$$\mathbf{e}_3 \times \mathbf{e}_1 = -\mathbf{e}_1 \times \mathbf{e}_3 = \mathbf{e}_2$$
$$\mathbf{e}_2 \times \mathbf{e}_3 = -\mathbf{e}_3 \times \mathbf{e}_2 = \mathbf{e}_1$$

and

$$\mathbf{e}_1 \times \mathbf{e}_1 = \mathbf{e}_2 \times \mathbf{e}_2 = \mathbf{e}_3 \times \mathbf{e}_3 = \mathbf{0}$$

These equations can all be expressed as

$$\mathbf{e}_i \times \mathbf{e}_j = \epsilon_{ijk} \mathbf{e}_k \tag{4.2}$$

Fundamentals of Continuum Mechanics, First Edition. John W. Rudnicki.
© 2015 John Wiley & Sons, Ltd. Published 2015 by John Wiley & Sons, Ltd.

where the *permutation symbol* is defined such that

$$\epsilon_{ijk} = \begin{cases} 0 & \text{if any two indices are equal} \\ +1 & \text{if } (ijk) \text{ is an even permutation of } (123), \text{i.e., } 123, 312, 231 \\ -1 & \text{if } (ijk) \text{ is an odd permutation of } (123), \text{i.e., } 213, 321, 132 \end{cases} \qquad (4.3)$$

Interchanging any two indices of ϵ_{ijk} changes an even permutation to odd and vice versa.

The relation (4.2) can be used to determine the component form of the cross product of two vectors in (4.1):

$$\mathbf{w} = u_i v_j (\mathbf{e}_i \times \mathbf{e}_j) = u_i v_j \epsilon_{ijk} \mathbf{e}_k \qquad (4.4)$$

Writing out the sum in (4.4) and using the properties of the permutation symbol (4.3) yields

$$\mathbf{w} = (u_2 v_3 - v_2 u_3)\mathbf{e}_1 + (u_3 v_1 - u_1 v_3)\mathbf{e}_2 + (u_1 v_2 - v_1 u_2)\mathbf{e}_3$$

This equation can be arranged as the determinant

$$\mathbf{w} = \begin{vmatrix} \mathbf{e}_1 & \mathbf{e}_2 & \mathbf{e}_3 \\ u_1 & u_2 & u_3 \\ v_1 & v_2 & v_3 \end{vmatrix}$$

4.1 Properties of the Cross Product

To illustrate manipulation of index notation and the permutation symbol, we confirm previously introduced properties of the cross product.

First, to show that reversing the order of the vectors introduces a minus sign we express the cross product in index notation:

$$\mathbf{u} \times \mathbf{v} = \epsilon_{ijk} u_i v_j \mathbf{e}_k \qquad (4.5)$$

$$= -\epsilon_{jik} u_i v_j \mathbf{e}_k \qquad (4.6)$$

$$= -\epsilon_{lmk} v_l u_m \mathbf{e}_k = -\mathbf{v} \times \mathbf{u} \qquad (4.7)$$

The second line introduces a minus sign because the order of the indices i and j in ϵ_{ijk} is reversed. In the third line, the indices i and j are simply relabeled (this can be done because they are dummy or summation indices) and this is recognized as the component form of $\mathbf{v} \times \mathbf{u}$.

Now we show that the cross product is orthogonal to each of the vectors in the product:

$$\mathbf{u} \cdot \mathbf{w} = \mathbf{w} \cdot \mathbf{u} = 0 \qquad (4.8)$$

where $\mathbf{w} = \mathbf{u} \times \mathbf{v}$. Substituting the expression for $\mathbf{w}$ in (4.8) and expressing in component form gives

$$\mathbf{u} \cdot (\mathbf{u} \times \mathbf{v}) = (u_i \mathbf{e}_i) \cdot (\epsilon_{klm} u_k v_l \mathbf{e}_m)$$

$$= u_i u_k v_l (\mathbf{e}_i \cdot \mathbf{e}_m) \epsilon_{klm}$$

$$= u_i u_k v_l \epsilon_{kli} = 0$$

Because the scalar product pertains to vectors, the expression can be regrouped as in the second line and carrying out the scalar product results in the third line. Interchanging two indices on ϵ_{lik} introduces a minus sign and relabeling the indices as in (4.6) and (4.7) shows that the expression is equal to its negative and, hence, must be zero. This is a particular case of the more general result of Example 3.6.3.

4.2 Triple Scalar Product

We have already noted that the *triple scalar product* $\mathbf{u} \times \mathbf{v} \cdot \mathbf{w}$ gives the volume of the parallelepiped with $\mathbf{u}$, $\mathbf{v}$, and $\mathbf{w}$ as edges (or the negative of the volume depending on the ordering of the vectors) (Figure 1.6). The component form follows from (4.4):

$$\mathbf{u} \times \mathbf{v} \cdot \mathbf{w} = \epsilon_{ijk} u_i v_j w_k \tag{4.9}$$

The triple scalar product can also be represented by the following determinant:

$$\mathbf{u} \times \mathbf{v} \cdot \mathbf{w} = \begin{vmatrix} u_1 & u_2 & u_3 \\ v_1 & v_2 & v_3 \\ w_1 & w_2 & w_3 \end{vmatrix} \tag{4.10}$$

where the right hand side of (4.9) indicates the operations implied by the determinant on the right side of (4.10). The next chapter will discuss determinants in more detail.

Because the triple scalar product vanishes if the vectors $\mathbf{u}$, $\mathbf{v}$, and $\mathbf{w}$ are coplanar, the condition is also expressed by the vanishing of this determinant. In this case, any one of the vectors can be expressed as a linear combination of the other two or, equivalently, one row or column of the matrix is a linear combination of the remaining two.

Replacing $\mathbf{u}$ by $\mathbf{e}_i = \delta_{ip} \mathbf{e}_p$ and similarly for $\mathbf{v}$ and $\mathbf{w}$ gives the triple scalar product of three orthonormal unit vectors

$$\mathbf{e}_i \cdot (\mathbf{e}_j \times \mathbf{e}_k) = \epsilon_{ijk} = \begin{vmatrix} \delta_{i1} & \delta_{i2} & \delta_{i3} \\ \delta_{j1} & \delta_{j2} & \delta_{j3} \\ \delta_{k1} & \delta_{k2} & \delta_{k3} \end{vmatrix} \tag{4.11}$$

The determinant is skew symmetric with respect to the interchange of (i, j, k), demonstrating that the interchange of rows implies multiplication by (-1). When $(i, j, k) = (123)$, the determinant equals one.

4.3 Triple Vector Product

The component form of the triple vector product introduced in (1.5) is

$$\mathbf{u} \times (\mathbf{v} \times \mathbf{w}) = u_j \mathbf{e}_j \times (\mathbf{e}_k \epsilon_{klm} v_l w_m)$$

$$= \mathbf{e}_i \epsilon_{ijk} \epsilon_{klm} u_j v_l w_m \tag{4.12}$$

The following ϵ–δ identity is essential for manipulating the component forms of expressions involving two vector products:

$$\epsilon_{ijk}\epsilon_{imn} = \delta_{jm}\delta_{kn} - \delta_{jn}\delta_{km} \tag{4.13}$$

Cyclically rotating the indices on the first $\epsilon_{ijk} = \epsilon_{kij}$ in (4.12) and relabeling indices puts this expression in the same form as (4.13). Using the ϵ–δ identity gives the component form, which can be recognized as the coordinate-free form in the second line below:

$$\mathbf{u} \times (\mathbf{v} \times \mathbf{w}) = (v_i\mathbf{e}_i)u_kw_k - (w_i\mathbf{e}_i)u_kv_k$$
$$= \mathbf{v}(\mathbf{u} \cdot \mathbf{w}) - \mathbf{w}(\mathbf{u} \cdot \mathbf{v}) \tag{4.14}$$

Thus, in (1.5) α is $\mathbf{u} \cdot \mathbf{w}$ and $\beta = -\mathbf{u} \cdot \mathbf{v}$.

4.4 Applications of the Cross Product

Two applications of the cross product familiar from basic mechanics represent the velocity due to rigid body rotation and the moment of a force about a point.

4.4.1 Velocity due to Rigid Body Rotation

In a rigid body the distance between any two points is fixed. Consider rotation of a rigid body with angular velocity ω about an axis designated by the unit vector $\mathbf{n}$, as shown in Figure 4.1. The angular velocity vector is

$$\boldsymbol{\omega} = \omega\mathbf{n}$$

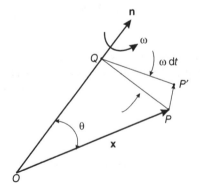

Figure 4.1   Velocity due to rigid body rotation.

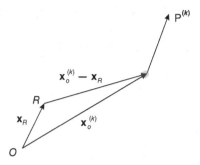

Figure 4.2   Moment of a force $\mathbf{P}^{(k)}$ on the k particle.

A point P, in the rigid body, is located by the position vector $\mathbf{x}$. The vector $\mathbf{n} \times \mathbf{x}$ is in direction PP' and has magnitude $|\mathbf{x}| \sin \theta$. But $|\mathbf{x}| \sin \theta = PQ$ is the perpendicular distance from P to the axis of rotation. Therefore, in time dt, the displacement is

$$\mathbf{du} = \omega \mathbf{n} \times \mathbf{x} \, dt$$

In the limit $dt \to 0$, the velocity is

$$\mathbf{v} = \boldsymbol{\omega} \times \mathbf{x}$$

4.4.2   Moment of a Force $\mathbf{P}$ about O

Consider a system of particles. The kth particle is shown in Figure 4.2. Each particle is located at $\mathbf{x}_o^{(k)}$ relative to a point O and subjected to a force $\mathbf{P}^{(k)}$. The moment of the force $\mathbf{P}^{(k)}$ about O is

$$\mathbf{M}_o^{(k)} = \mathbf{x}_o^{(k)} \times \mathbf{P}^{(k)}, \quad \text{(no sum on } k\text{)} \tag{4.15}$$

For k particles in equilibrium, the sum of the forces must vanish

$$\sum_k \mathbf{P}^{(k)} = \mathbf{0} \tag{4.16}$$

and the sum of the moments must vanish

$$\sum_k \mathbf{x}_o^{(k)} \times \mathbf{P}^{(k)} = \mathbf{0} \tag{4.17}$$

A result from statics states that if the sum of the moments about one point vanishes for a system of particles in equilibrium, then the sum of the moments about any point vanishes.

Consider another point R where $\mathbf{x}_R$ is the vector from the origin to R. Since (4.17) is satisfied,

$$\sum_k \left\{ \left(\mathbf{x}_o^{(k)} - \mathbf{x}_R \right) \times \mathbf{P}^{(k)} + \mathbf{x}_R \times \mathbf{P}^{(k)} \right\} = \mathbf{0}$$

$$\sum_k \left(\mathbf{x}_o^{(k)} - \mathbf{x}_R \right) \times \mathbf{P}^{(k)} + \mathbf{x}_R \times \sum_k \mathbf{P}^{(k)} = \mathbf{0}$$

But the last term vanishes because of (4.16) and hence the sum of the moments about R must vanish.

4.5   Non-orthonormal Basis

Although we use almost exclusively orthogonal unit vectors as base vectors, it is not necessary and any three non-coplanar vectors can be used as a basis. This section and Problems 4.13 and 4.14 give a brief introduction to the use of non-orthonormal base vectors.

For an orthonormal basis, an arbitrary vector $\mathbf{u}$ is given by (3.5). The coefficient of each basis vector in (3.5) is also the projection of the vector on the base vector (3.4). This is a special feature of rectangular Cartesian systems.

Let $\mathbf{g}_1$, $\mathbf{g}_2$, and $\mathbf{g}_3$ be three arbitrary non-coplanar vectors that we choose as a basis. Then an arbitrary vector $\mathbf{u}$ is given by

$$\mathbf{u} = u^i \mathbf{g}_i$$

where the summation convention applies for repeated indices regardless of whether they are superscripts or subscripts (the reason for using superscripts will become clear). Note that the component u^i is not given by the scalar product of $\mathbf{u}$ with $\mathbf{g}_i$, $u^i \neq \mathbf{g}_i \cdot \mathbf{u}$, because $\mathbf{g}_i \cdot \mathbf{g}_j \neq \delta_{ij}$. Instead u^i is given by the scalar product of $\mathbf{u}$ with the *dual* or *reciprocal* base vectors $\mathbf{g}^i$ defined by

$$\mathbf{g}_i \cdot \mathbf{g}^j = \delta_{ij}$$

Because $\mathbf{g}^1 \cdot \mathbf{g}_2 = 0$ and $\mathbf{g}^1 \cdot \mathbf{g}_3 = 0$, $\mathbf{g}^1$ is orthogonal to $\mathbf{g}_2$ and $\mathbf{g}_3$ and must be proportional to the cross product of $\mathbf{g}_2$ and $\mathbf{g}_3$:

$$\mathbf{g}^1 = g^{-1} \mathbf{g}_1 \times \mathbf{g}_2$$

Forming $\mathbf{g}^1 \cdot \mathbf{g}_1$ gives $g = \mathbf{g}_1 \cdot \mathbf{g}_2 \times \mathbf{g}_3$ where $g \neq 0$ because the vectors are non coplanar (see Problem 1.11). Thus the vectors of the reciprocal basis are given by

$$\mathbf{g}^1 = g^{-1} \mathbf{g}_2 \times \mathbf{g}_3, \quad \mathbf{g}^2 = g^{-1} \mathbf{g}_3 \times \mathbf{g}_1, \quad \mathbf{g}^3 = g^{-1} \mathbf{g}_1 \times \mathbf{g}_2 \qquad (4.18)$$

Thus, any arbitrary vector $\mathbf{u}$ can be expressed as

$$\mathbf{u} = (\mathbf{u} \cdot \mathbf{g}^1)\mathbf{g}_1 + (\mathbf{u} \cdot \mathbf{g}^2)\mathbf{g}_2 + (\mathbf{u} \cdot \mathbf{g}^3)\mathbf{g}_3$$

Substituting from (4.18) and using Problem 1.2 yield

$$\mathbf{u} = (\mathbf{u} \times \mathbf{g}_2) \cdot \mathbf{g}_3 \mathbf{g}_1 g^{-1} + (\mathbf{g}_1 \times \mathbf{u}) \cdot \mathbf{g}_3 \mathbf{g}_2 g^{-1} + (\mathbf{g}_1 \times \mathbf{g}_2) \cdot \mathbf{u} \mathbf{g}_3 g^{-1}$$

4.6 Example

Use index notation to prove

$$(\mathbf{a} \times \mathbf{b}) \cdot (\mathbf{a} \times \mathbf{b}) = a^2 b^2 \sin^2 \theta$$

where θ is the angle between $\mathbf{a}$ and $\mathbf{b}$.

Here

$$|\mathbf{a} \times \mathbf{b}| = (\mathbf{a} \times \mathbf{b}) \cdot (\mathbf{a} \times \mathbf{b})$$
$$= \epsilon_{ijk}\mathbf{e}_i a_j b_k \cdot \epsilon_{pqr}\mathbf{e}_p a_q b_r$$
$$= (\mathbf{e}_i \cdot \mathbf{e}_p)\epsilon_{ijk}\epsilon_{pqr}a_j b_k a_q b_r$$
$$= \epsilon_{ijk}\epsilon_{iqr}a_j b_k a_q b_r$$

Using the ϵ–δ identity (4.13),

$$|\mathbf{a} \times \mathbf{b}| = (\delta_{jq}\delta_{kr} - \delta_{jr}\delta_{kq})a_j b_k a_q b_r$$
$$= (a_j a_j)(b_k b_k) - (a_j b_j)(a_k b_k)$$
$$= (\mathbf{a} \cdot \mathbf{a})(\mathbf{b} \cdot \mathbf{b}) - (\mathbf{a} \cdot \mathbf{b})(\mathbf{a} \cdot \mathbf{b})$$

Using (1.2) gives

$$|\mathbf{a} \times \mathbf{b}| = a^2 b^2 - a^2 b^2 \cos^2 \theta$$
$$= a^2 b^2 \sin^2 \theta$$

Exercises

4.1 Beginning with (4.2), show that

$$\epsilon_{ijm} = \mathbf{e}_m \cdot (\mathbf{e}_i \times \mathbf{e}_j)$$

and explain why no parentheses are needed.

4.2 Construct a proof of the ϵ–δ identity (4.13) by noting that each of the four free indices j, k, m, n can take on only three values: 1, 2, or 3. Therefore at least two indices must be identical and as a result it is possible to enumerate the various outcomes.

4.3 **(a)** Show that contracting two of the indices of the ϵ–δ identity gives

$$\epsilon_{pqi}\epsilon_{pqj} = 2\delta_{ij}$$

(b) Show that contracting all three indices gives

$$\epsilon_{pqr}\epsilon_{pqr} = 6$$

4.4   Show that

$$\mathbf{e}_m = \frac{1}{2}\epsilon_{mij}\mathbf{e}_i \times \mathbf{e}_j$$

where the $\mathbf{e}_m$ are (right-handed) orthonormal base vectors.

4.5   We showed that the triple vector product $\mathbf{u} \times (\mathbf{v} \times \mathbf{w})$ has the representation (4.14).
   (a)   Show that the result can be written as

$$\mathbf{u} \times (\mathbf{v} \times \mathbf{w}) = \mathbf{u} \cdot (\mathbf{wv} - \mathbf{vw})$$

   (b)   Show that (4.14) implies

$$\mathbf{u} \times (\mathbf{v} \times \mathbf{w}) + \mathbf{v} \times (\mathbf{w} \times \mathbf{u}) + \mathbf{w} \times (\mathbf{u} \times \mathbf{v}) = 0$$

4.6   Determine the scalars α and β in the relation

$$(\mathbf{u} \times \mathbf{v}) \times \mathbf{w} = \alpha\mathbf{u} + \beta\mathbf{v}$$

where $\mathbf{u}$, $\mathbf{v}$, and $\mathbf{w}$ are vectors.

4.7   If $\mathbf{n}$ is a unit vector show that any vector $\mathbf{u}$ can be written as the sum of terms parallel and perpendicular to $\mathbf{n}$:

$$\mathbf{u} = \mathbf{n}(\mathbf{n} \cdot \mathbf{u}) + \mathbf{n} \times (\mathbf{u} \times \mathbf{n})$$

4.8   Use index notation to prove the following relations involving the cross products of vectors $\mathbf{a}$, $\mathbf{b}$, $\mathbf{c}$, and $\mathbf{d}$:
   (a) $(\mathbf{a} \times \mathbf{b}) \cdot (\mathbf{c} \times \mathbf{d}) = (\mathbf{a} \cdot \mathbf{c})(\mathbf{b} \cdot \mathbf{d}) - (\mathbf{a} \cdot \mathbf{d})(\mathbf{b} \cdot \mathbf{c})$
   (b) $(\mathbf{a} \times \mathbf{b}) \times (\mathbf{c} \times \mathbf{d}) = [\mathbf{c} \cdot (\mathbf{d} \times \mathbf{a})]\mathbf{b} - [\mathbf{c} \cdot (\mathbf{d} \times \mathbf{b})]\mathbf{a}$

4.9   For a skew-symmetric tensor $\mathbf{W}$ the following product

$$\mathbf{W} \cdot \mathbf{u} = \mathbf{w} \times \mathbf{u}$$

for an arbitrary vector $\mathbf{u}$ defines the vector $\mathbf{w}$. Determine the components of $\mathbf{W}$ in terms of the components of $\mathbf{w}$ and those of $\mathbf{w}$ in terms of the components of $\mathbf{W}$. (What happens if $\mathbf{W}$ is not skew symmetric?)

4.10   If $\mathbf{n}$ is a unit vector $\mathbf{n} \cdot \mathbf{n} = 1$. Differentiating gives $\dot{\mathbf{n}} \cdot \mathbf{n} = 0$. Consequently $\dot{\mathbf{n}}$ is orthogonal to $\mathbf{n}$ and can be represented as $\mathbf{w} \times \mathbf{n}$.
   (a)   Show that

$$\mathbf{w} = \mathbf{n} \times \dot{\mathbf{n}}$$

   (b)   Use the result of Problem 4.9 to show that

$$\mathbf{W} = \dot{\mathbf{n}}\mathbf{n} - \mathbf{n}\dot{\mathbf{n}}$$

4.11 Use the Cartesian component forms of the vectors u and v and the tensor **F** to show that

$$\mathbf{u} \cdot (\mathbf{v} \times \mathbf{F}) = (\mathbf{u} \times \mathbf{v}) \cdot \mathbf{F}$$

4.12 The adjugate **F*** of a tensor **F** is defined by the following relation for all vectors **u** and **v**:

$$(\mathbf{u} \times \mathbf{v}) \cdot \mathbf{F}^* = (\mathbf{F} \cdot \mathbf{u}) \times (\mathbf{F} \cdot \mathbf{v})$$

Show that the rectangular Cartesian components of **F*** are given by

$$F^*_{ni} = \frac{1}{2} \epsilon_{iqr} \epsilon_{nkl} F_{qk} F_{rl}$$

4.13 Determine the dual base vectors if $\mathbf{g}_1 = \mathbf{e}_2 + \mathbf{e}_3$, $\mathbf{g}_2 = \mathbf{e}_1 + \mathbf{e}_3$, and $\mathbf{g}_3 = \mathbf{e}_1 + \mathbf{e}_2$.

4.14 (a) If $\mathbf{g}_1 = 3\mathbf{e}_1 + \mathbf{e}_2$ and $\mathbf{g}_2 = -3\mathbf{e}_1 + \mathbf{e}_2$ are two vectors to be used as a basis in the 1–2 plane, construct the dual base vectors and show them in relation to the original base vectors in a graph.

(b) Express the vector $\mathbf{v} = \mathbf{e}_1 + 5\mathbf{e}_2$ in terms of the original and dual bases from (a) and show them in a diagram.

5

Determinants

Expressions involving the determinants of matrices will arise frequently in later chapters. Consequently, it is useful to review some of their properties. In addition, this will provide further illustrations of manipulating the permutation symbol.

As noted in Chapter 4, the component form of the triple scalar product can be represented with the permutation symbol (4.9) or as a determinant (4.10). This correspondence suggests that ϵ_{ijk} can be useful in representing determinants more generally. For example, if we replace the components of the vector $\mathbf{u}$, namely, u_1, u_2, and u_3, by M_{11}, M_{12}, and M_{13}, and similarly replace the components of $\mathbf{v}$ and $\mathbf{w}$ by M_{2i} and M_{3i}, then (4.10) becomes an expression for the determinant of the matrix M with components M_{ij}:

$$\det(M) = \epsilon_{ijk} M_{1i} M_{2j} M_{3k} = \begin{vmatrix} M_{11} & M_{12} & M_{13} \\ M_{21} & M_{22} & M_{23} \\ M_{31} & M_{32} & M_{33} \end{vmatrix} \tag{5.1}$$

Writing out the summation and using the properties of the permutation symbol (4.3) gives

$$\det(M) = M_{11}(M_{22}M_{33} - M_{23}M_{32}) - M_{12}(M_{21}M_{33} - M_{23}M_{31})$$
$$+ M_{13}(M_{21}M_{32} - M_{22}M_{31})$$
$$= M_{11} \begin{vmatrix} M_{22} & M_{23} \\ M_{32} & M_{33} \end{vmatrix} - M_{12} \begin{vmatrix} M_{21} & M_{23} \\ M_{31} & M_{33} \end{vmatrix} + M_{13} \begin{vmatrix} M_{21} & M_{22} \\ M_{31} & M_{32} \end{vmatrix}$$

where the second equality results from arranging the coefficients of M_{11}, M_{12}, and M_{13} as 2×2 determinants. The sign of the coefficient is $(-1)^{i+j}$ where i and j are the row and column numbers. Thus, the summation represents an expansion of the determinant by the first row. The signed coefficients of M_{11}, M_{12}, and M_{13} are called the *cofactors* of these terms. Note that each term has one and only one element from each row and column.

Fundamentals of Continuum Mechanics, First Edition. John W. Rudnicki.
© 2015 John Wiley & Sons, Ltd. Published 2015 by John Wiley & Sons, Ltd.

Alternatively, we could replace the components of $\mathbf{u}$ by M_{11}, M_{21}, and M_{31}, and similarly replace the components of $\mathbf{v}$ and $\mathbf{w}$ by M_{i2} and M_{i3}. This leads to an expansion about the first column:

$$\det(M) = \epsilon_{ijk} M_{i1} M_{j2} M_{k3} \tag{5.2}$$

Consequently, the determinant of a matrix and its transpose are identical:

$$\det(M) = \det(M^T)$$

Because the appearance of the specific numbers 1, 2, and 3 in (5.1) and (5.2) is sometimes inconvenient, it is useful to develop another expression in terms of arbitrary indices. To this end, consider the quantity

$$h_{lmn} = \epsilon_{ijk} M_{li} M_{mj} M_{nk}$$

First, note from (5.1) that $h_{123} = \det M$. The argument of equations (4.5) to (4.7) shows that h_{lmn} is skew symmetric with respect to the interchange of (l, m, n) and zero if any two indices are the same. From these results, we can conclude that

$$\epsilon_{ijk} M_{li} M_{mj} M_{nk} = \epsilon_{lmn} \det(M) \tag{5.3}$$

Similarly, beginning with (5.2) we can show that

$$\epsilon_{lmn} M_{li} M_{mj} M_{nk} = \epsilon_{ijk} \det(M) \tag{5.4}$$

5.1   Cofactor

Writing the determinant as

$$\det(M) = M_{1i} c_{1i}$$

identifies

$$c_{1i} = \epsilon_{ijk} M_{2j} M_{3k} \tag{5.5}$$

as the *cofactor* of M_{1i}. It is not necessary to expand about the first row (5.1) or first column (5.2). We can expand about any row or column

$$\det(M) = M_{pi} c_{pi} = M_{ip} c_{ip} \text{ (no sum on } p\text{)}$$

(Because we have adopted the convention that a repeated index implies summation, we must explicitly indicate here that p is not to be summed. If p is summed, the result is $3 \det(M)$.)

Similarly to (5.3) and (5.4), we can derive an expression for the cofactor that does not include specific values of the indices. The cofactor of M_{1i} in (5.1) is (5.5). To rewrite this in a more general way for arbitrary indices we first multiply the right side by $\epsilon_{123} = 1$ and write it as the sum of two terms

$$c_{1i} = \frac{1}{2}\epsilon_{123}\epsilon_{ijk}M_{2j}M_{3k} + \frac{1}{2}\epsilon_{123}\epsilon_{ijk}M_{2j}M_{3k}$$

$$= \frac{1}{2}\epsilon_{123}\epsilon_{ijk}M_{2j}M_{3k} + \frac{1}{2}\epsilon_{132}\epsilon_{ijk}M_{3j}M_{2k}$$

$$= \frac{1}{2}\epsilon_{ijk}\epsilon_{1mn}M_{mj}M_{nk}$$

In the second term, reversing 3 and 2 in ϵ_{123} and j and k in ϵ_{ijk} gives the second line. The third line follows from noting that 2 and 3 are the only subscripts giving nonzero terms in ϵ_{1mn}. Because this expression applies for any value of the first index we can write

$$c_{li} = \frac{1}{2}\epsilon_{lmn}\epsilon_{ijk}M_{mj}M_{nk} \tag{5.6}$$

The transpose of this matrix is the *adjugate* of M (see Problem 4.12)

$$M^*_{li} = \frac{1}{2}\epsilon_{imn}\epsilon_{ljk}M_{mj}M_{nk} \tag{5.7}$$

5.2   Inverse

The adjugate is related to the inverse of a matrix. Multiplying the adjugate (5.7) by M_{pi} gives

$$M_{pi}M^*_{li} = \frac{1}{2}\epsilon_{imn}(\epsilon_{ljk}M_{pi}M_{mj}M_{nk})$$

We can rewrite the term in parentheses on the right side by using (5.3) and the ϵ–δ identity (4.13) to give

$$M_{pi}M^*_{il} = \frac{1}{2}\epsilon_{lmn}\epsilon_{pmn}\det(M) = \delta_{pl}\det(M)$$

Dividing both sides by $\det(M)$ gives

$$M_{pi}\frac{M^*_{il}}{\det(M)} = \delta_{pl}$$

If the right side is arranged as a matrix, it is the identity, i.e., the matrix with ones on the diagonal and zeros elsewhere. Consequently, the term multiplying M_{pi} must be an expression for the inverse of this matrix. Therefore, the inverse is given by

$$M^{-1}_{il} = \frac{M^*_{il}}{\det(M)} \tag{5.8}$$

Note that if $\det(M) = 0$, the inverse will not exist. Recall that when the determinant is interpreted as the triple scalar product of three vectors, it vanishes if the three vectors are coplanar. In other words, the third vector can be expressed in terms of a linear combination of the other two or, equivalently, one row of the matrix is a linear combination of the remaining two.

5.3 Example

Prove that the product of two determinants is the determinant of the product matrix

$$\det(A)\det(B) = \det(C)$$

where $C_{kl} = A_{kp}B_{pl}$.
Using (5.1) gives

$$\det(A)\det(B) = \det(A)\epsilon_{mnp}b_{m1}b_{n2}b_{p3}$$

Then using (5.3) gives

$$\det(A)\det(B) = \epsilon_{ijk}(a_{im}b_{m1})(a_{jn}b_{n2})(a_{kp}b_{p3})$$

Thus the left hand side is $\det(C)$.

Exercises

5.1 Write out equation (5.2) and verify that this does represent an expansion in columns.

5.2 Use equation (5.1) or (5.2) to show that interchanging two rows or columns changes the sign of a determinant.

5.3 Derive equation (5.4).

5.4 Show that the same expression for the cofactor, equation (5.6), results if one begins with the column expansion (5.2).

5.5 Verify equation (5.6) by writing out c_{21} and c_{13}.

5.6 Use equation (4.10) and the result of Example 5.3 to show that

$$(\mathbf{a} \cdot \mathbf{b} \times \mathbf{c})(\mathbf{d} \cdot \mathbf{e} \times \mathbf{f}) = \begin{vmatrix} \mathbf{a} \cdot \mathbf{d} & \mathbf{a} \cdot \mathbf{e} & \mathbf{a} \cdot \mathbf{f} \\ \mathbf{b} \cdot \mathbf{d} & \mathbf{b} \cdot \mathbf{e} & \mathbf{b} \cdot \mathbf{f} \\ \mathbf{c} \cdot \mathbf{d} & \mathbf{c} \cdot \mathbf{e} & \mathbf{c} \cdot \mathbf{f} \end{vmatrix}$$

5.7 Use equation (4.11) and the results of Example 5.3 and Problem 5.6 to show that

$$\epsilon_{ijk}\epsilon_{mnp} = \begin{vmatrix} \delta_{i1} & \delta_{i2} & \delta_{i3} \\ \delta_{j1} & \delta_{j2} & \delta_{j3} \\ \delta_{k1} & \delta_{k2} & \delta_{k3} \end{vmatrix} \begin{vmatrix} \delta_{m1} & \delta_{m2} & \delta_{m3} \\ \delta_{n1} & \delta_{n2} & \delta_{n3} \\ \delta_{p1} & \delta_{p2} & \delta_{p3} \end{vmatrix}$$

$$= \begin{vmatrix} \delta_{im} & \delta_{in} & \delta_{ip} \\ \delta_{jm} & \delta_{jn} & \delta_{jp} \\ \delta_{km} & \delta_{kn} & \delta_{kp} \end{vmatrix}$$

Verify that setting $i = m$ gives the ϵ–δ identity (4.13).

5.8 Show that

$$\det(M) = \frac{1}{6}\epsilon_{ijk}\epsilon_{lmn}M_{il}M_{jm}M_{kn}$$

5.9 Use index notation to prove the following identity: $[(\mathbf{M} \cdot \mathbf{a}) \times (\mathbf{M} \cdot \mathbf{b})] \cdot \mathbf{M} = \det(\mathbf{M})(\mathbf{a} \times \mathbf{b})$.

5.10 Show that

$$(\mathbf{M} \cdot \mathbf{a}) \cdot (\mathbf{M} \cdot \mathbf{b}) \times (\mathbf{M} \cdot \mathbf{c}) = \det(\mathbf{M})(\mathbf{a} \cdot \mathbf{b} \times \mathbf{c})$$

5.11 Use Problem 5.10 to show that the determinant of an orthogonal tensor is ± 1 where the positive (negative) sign applies if the rotation is from right-handed to right-(left) handed.

6

Change of Orthonormal Basis

In Chapter 1 and Chapter 2 vectors and tensors were introduced in coordinate-free form. This representation was motivated by recognizing that the quantities we use vectors to represent are physical entities and hence independent of the coordinate system used to describe them. In practice, it is, however, necessary to refer tensors and vectors to a coordinate system. In Chapter 3 we introduced components relative to rectangular Cartesian coordinate systems described by unit, orthogonal (i.e., orthonormal) base vectors. This is a convenient simplification. Nevertheless, we can retain a version of the coordinate-free concept by requiring that the components of tensors and vectors be expressed in terms of *any* rectangular Cartesian coordinate system. In this chapter we will show that this requirement imposes certain relations between the components in different systems. These relations provide an alternative approach to defining vectors and tensors.

Consider the two coordinate systems shown in Figure 6.1: the 123 system with base vectors $\mathbf{e}_1, \mathbf{e}_2, \mathbf{e}_3$ and the $1'2'3'$ system with base vectors $\mathbf{e}'_1, \mathbf{e}'_2, \mathbf{e}'_3$. The base vectors in the primed and unprimed systems are related by

$$\mathbf{e}'_j = \mathbf{A} \cdot \mathbf{e}_j \tag{6.1}$$

Because both the primed and unprimed systems of base vectors are orthonormal, $\mathbf{A}$ is an orthogonal tensor (2.12). Forming the scalar product with $\mathbf{e}_i$ in (6.1) gives

$$\mathbf{e}_i \cdot \mathbf{e}'_j = \cos(i, j') = \mathbf{e}_i \cdot \mathbf{A} \cdot \mathbf{e}_j = A_{ij} \tag{6.2}$$

where $\cos(i, j')$ is the cosine of the angle between the i axis and the j' axis. Thus, in the component A_{ij}, the second subscript (j in this case) is associated with the primed system. Either (6.1) or (6.2) leads to the dyadic representation

$$\mathbf{A} = \mathbf{e}'_k \mathbf{e}_k$$

Fundamentals of Continuum Mechanics, First Edition. John W. Rudnicki.
© 2015 John Wiley & Sons, Ltd. Published 2015 by John Wiley & Sons, Ltd.

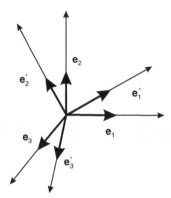

Figure 6.1   Rotation of the base vectors $\mathbf{e}_i$ to a new system $\mathbf{e}_i'$.

Because the inverse of an orthogonal tensor is equal to its transpose (2.14), the unprimed base vectors can be given in terms of the primed ones by

$$\mathbf{e}_m = \mathbf{A}^T \cdot \mathbf{e}_m' \qquad (6.3)$$

and

$$\mathbf{e}_n' \cdot \mathbf{e}_m = \cos(m, n') = \mathbf{e}_n' \cdot \mathbf{A}^T \cdot \mathbf{e}_m' = A_{nm}^T = A_{mn}$$

which agrees with (6.2). These properties reinforce the choice of the name *orthogonal* for this type of tensor: it rotates one system of orthogonal unit vectors into another system of orthogonal unit vectors. Equation (2.14) is expressed in index form as

$$A_{ik}A_{jk} = A_{ki}A_{kj} = \delta_{ij} \qquad (6.4)$$

Equation (6.4) reflects the following relation in terms of the direction cosines:

$$\sum_{k'=1}^{3} \cos(i, k') \cos(j, k') = \sum_{k=1}^{3} \cos(k, i') \cos(k, j') = \delta_{ij}$$

6.1   Change of Vector Components

Now consider a vector $\mathbf{v}$. Because $\mathbf{v}$ represents a physical entity, we can express it in terms of components in either system

$$\mathbf{v} = v_i \mathbf{e}_i = v_j' \mathbf{e}_j'$$

Both the v_i and the v_j' represent the *same* vector; they simply furnish different descriptions. Equations (6.1) and (6.3) between the base vectors impose relations between the v_i and v_j'. The component in the primed system is obtained by forming the scalar product of **v** with the base vector in the primed system:

$$
\begin{aligned}
v'_k &= \mathbf{e}'_k \cdot \mathbf{v} = \mathbf{e}'_k \cdot (v_i \mathbf{e}_i) \\
&= v_i \mathbf{e}'_k \cdot \mathbf{e}_i \\
&= v_i A_{ik}
\end{aligned}
\tag{6.5}
$$

Equations (6.5) can be represented as a matrix equation

$$
\begin{bmatrix} v'_1 & v'_2 & v'_3 \end{bmatrix} = \begin{bmatrix} v_1 & v_2 & v_3 \end{bmatrix} \begin{bmatrix} A_{11} & A_{12} & A_{13} \\ A_{21} & A_{22} & A_{23} \\ A_{31} & A_{32} & A_{33} \end{bmatrix}
$$

or, alternatively, as

$$
\begin{bmatrix} v'_1 \\ v'_2 \\ v'_3 \end{bmatrix} = \begin{bmatrix} A_{11} & A_{21} & A_{31} \\ A_{12} & A_{22} & A_{32} \\ A_{13} & A_{23} & A_{33} \end{bmatrix} \begin{bmatrix} v_1 \\ v_2 \\ v_3 \end{bmatrix}
\tag{6.6}
$$

Similarly, the components of **v** in the unprimed system can be expressed in terms of the components in the primed system

$$
v_i = A_{ik} v'_k
\tag{6.7}
$$

or in matrix form

$$
[v] = [A][v']
\tag{6.8}
$$

Note that according to (6.1) the tensor **A** rotates the unprimed base vectors into the primed base vectors, but it is the components of $\mathbf{A}^T$ that appear in the matrix equation (6.6) as implied by the index form (6.5).

We reiterate that (6.8) represents a relation between the components of the same vector. In the equation

$$
\mathbf{u} = \mathbf{A} \cdot \mathbf{v}'
$$

u and **v**′ are different vectors although the components of **u** in the primed system are equal to the components of **v** in the unprimed system. Writing

$$
\mathbf{v} = \mathbf{A} \cdot \mathbf{v}'
$$

should be avoided because it implies that **v** and **v**′ are different vectors and this notation can cause confusion between representations of a single vector expressed in terms of components in the primed and unprimed systems.

6.2 Definition of a Vector

Previously, we noted that vectors are directed line segments that add according to the parallelogram rule (Figure 1.2). This property of addition reflects the nature of the physical quantities that we represent as vectors, e.g., velocity and force. We now give another definition of a vector. This definition expresses the requirement that the quantities represented by vectors are physical entities that cannot depend on the coordinate systems used to represent them. A (Cartesian) vector $\mathbf{v}$ in three dimensions is a quantity with three components v_1, v_2, v_3 in one rectangular Cartesian system $\mathbf{e}_1\mathbf{e}_2\mathbf{e}_3$, which, under rotation of the coordinates to another Cartesian system $\mathbf{e}'_1\mathbf{e}'_2\mathbf{e}'_3$ (Figure 6.1), become components v'_1, v'_2, v'_3 with

$$v'_i = A_{ji}v_j \tag{6.9}$$

where A_{ji} is given by (6.2). This definition can then be used to deduce other properties of vectors. For example, we can show that the sum of two vectors is indeed a vector. If u_i and v_i are components of vectors then the three quantities $t_i = u_i + v_i$ are components of a vector $\mathbf{t}$ because they transform like one, according to the rule (6.9):

$$t'_i = u'_i + v'_i = A_{ji}u_j + A_{ji}v_j$$
$$= A_{ji}(u_j + v_j) = A_{ji}t_j$$

6.3 Change of Tensor Components

Expressions for the components of a tensor $\mathbf{F}$ with respect to a different set of base vectors, say $\mathbf{e}'_k$, follow from the relations for vector components:

$$v_k = F_{kl}u_l \tag{6.10}$$

and

$$v'_k = A_{mk}F_{mn}u_n$$
$$= A_{mk}F_{mn}A_{nl}u'_l$$
$$= F'_{kl}u'_l$$

Because this result applies for components of all vectors $\mathbf{u}$ and $\mathbf{v}$

$$F'_{kl} = A_{mk}F_{mn}A_{nl} \tag{6.11}$$

where, as before, A_{mk} is given by (6.2). Equation (6.11) can be written in matrix form as

$$[F'] = \begin{bmatrix} A_{11} & A_{21} & A_{31} \\ A_{12} & A_{22} & A_{32} \\ A_{13} & A_{23} & A_{33} \end{bmatrix} \begin{bmatrix} F_{11} & F_{12} & F_{13} \\ F_{21} & F_{22} & F_{23} \\ F_{31} & F_{32} & F_{33} \end{bmatrix} \begin{bmatrix} A_{11} & A_{12} & A_{13} \\ A_{21} & A_{22} & A_{23} \\ A_{31} & A_{32} & A_{33} \end{bmatrix}$$

or, more compactly, as

$$[F'] = [A]^T[F][A]$$

Similarly, the inversion is given by

$$F_{ij} = A_{il}A_{jk}F'_{lk} \tag{6.12}$$

or

$$[F] = [A][F'][A]^T$$

The relations between components of a tensor in different orthogonal coordinate systems can be used as a second definition of a tensor that is analogous to the definition of a vector: In any rectangular coordinate system, a tensor is defined by nine components that transform according to the rule (6.11) when the relation between unit base vectors is (6.1).

The relations (6.11) and (6.12) can be used to test whether the nine elements in an array are components of a tensor.

6.4   Isotropic Tensors

In later sections we will have occasion to work with a special class of tensors called isotropic tensors. They have the same components in every orthonormal coordinate system (see Aris, 1989, Sec. 2.7, pp. 30–34). All scalars (tensors of order 0) are isotropic. Equation (6.5) demonstrates that no vectors (except the null vector) are isotropic. For second-order tensors, the components in different rectangular coordinate systems are related by (6.11). For an isotropic second-order tensor, $F'_{ij} = F_{ij}$ and, hence,

$$F_{ij} = A_{ki}A_{lj}F_{kl} \tag{6.13}$$

for all A_{ki}. It is straightforward to verify that (6.13) is satisfied by any tensor of the form

$$F_{ij} = \alpha\delta_{ij} \tag{6.14}$$

where α is a scalar. Substituting (6.14) into the right side of (6.13) gives

$$F_{ij} = \alpha A_{ki}A_{kj}$$

and using (6.4) establishes (6.13).

This demonstrates that the identity tensor multiplied by a scalar is an isotropic tensor but does not answer the question of whether all isotropic tensors of second order must have this form. To do this, we again use (6.13). If this equation must be satisfied for all A_{ki} then it must certainly be satisfied for particular choices of the A_{ki}. Judicious choice of the A_{ki} can be used to demonstrate that all isotropic second-order tensors must have the form (6.14).

First, consider the transformation for which $A_{13} = A_{21} = A_{32} = 1$ are the only nonzero A_{ki}. This rotates the $\mathbf{e}_1$ into $\mathbf{e}_3'$, $\mathbf{e}_2$, into $\mathbf{e}_1'$ and $\mathbf{e}_3$ into $\mathbf{e}_2'$. Substituting into (6.13) gives

$$F_{11} = A_{i1}A_{j1}F_{ij} = F_{22}$$

and, similarly, $F_{22} = F_{33}$. Thus, the three diagonal components of F_{ij} must be identical: $F_{11} = F_{22} = F_{33} = \alpha$. Similarly, for the off-diagonal components

$$F_{12} = A_{i1}A_{j2}F_{ij} = A_{21}A_{32}F_{23} = F_{23}$$

Thus, the off-diagonal components must also be identical

$$F_{12} = F_{21} = F_{31} = F_{13} = F_{23} = F_{32} = \beta$$

Now consider the transformation corresponding to a rotation of $90°$ about the x_3 axis so that $A_{12} = -1 = -A_{21} = -A_{33}$ are the only nonzero A_{ki}. Applying (6.13) to F_{12} gives $F_{12} = A_{21}A_{12}F_{12} = -F_{12}$. Therefore $\beta = 0$ and (6.14) is the only isotropic tensor of order 2. A similar analysis can be used to show that the only isotropic tensor of third order is $\alpha \epsilon_{ijk}$.

Tensor products of isotropic tensors are also isotropic. Therefore, fourth-order tensors with components proportional to $\delta_{ij}\delta_{kl}$ are isotropic. In fact, all isotropic tensors of even order are sums and products of δ_{ij}. The number of possible terms for a tensor of order N is given by the combinatorial formula

$$\frac{N!}{2^{(N/2)}(N/2)!}$$

where $N!$ is the total number of ordered combinations, $(N/2)!$ is the number of ordered ways in which the pairs can be arranged, e.g., $\delta_{ij}\delta_{kl} = \delta_{kl}\delta_{ij}$, and $2^{(N/2)}$ accounts for the switching of indices of each pair, e.g., $\delta_{ij} = \delta_{ji}$. Applying this formula for $N = 4$ yields three possible combinations. Thus, the only isotropic tensor of fourth order has the form

$$V_{ijkl} = a\delta_{ij}\delta_{kl} + b\delta_{ik}\delta_{jl} + c\delta_{il}\delta_{jk} \tag{6.15}$$

Replacing b by $\bar{b} + \bar{c}$ and c by $\bar{b} - \bar{c}$ in (6.15) gives

$$V_{ijkl} = a\delta_{ij}\delta_{kl} + \bar{b}(\delta_{ik}\delta_{jl} + \delta_{il}\delta_{jk}) + \bar{c}(\delta_{ik}\delta_{jl} - \delta_{il}\delta_{jk}) \tag{6.16}$$

If $V_{ijkl} = V_{jikl}$ or $V_{ijkl} = V_{ijlk}$, $\bar{c} = 0$ and only two parameters are needed to define an isotropic fourth-order tensor with this additional symmetry. Another method of establishing this result will be given in Chapter 23.

6.5   Example

If F_{ij} and G_{ij} are components of second-order tensors show that the H_{ij} given by

$$F_{ik}H_{kj} = G_{ij}$$

are components of a second-order tensor.

Using (6.12) for F_{ij} and G_{ij} gives

$$A_{im}F'_{mn}A_{kn}H_{kj} = A_{ip}G'_{pq}A_{jq}$$

Multiplying both sides by $A_{ir}A_{js}$ and using (6.4) gives

$$F'_{rn}(A_{kn}H_{kj}A_{js}) = G'_{rs}$$

This identifies H'_{ns} as the term in parentheses. Consequently, the elements of the array H_{ij} transform according to (6.12) and hence are components of a tensor.

Exercises

6.1 Show that

$$v'_k = \mathbf{e}_k \cdot (\mathbf{A}^T \cdot \mathbf{v})$$

6.2 Derive equation (6.7).

6.3 Derive equation (6.12).

6.4 Verify that equation (6.4) is the component representation of equation (2.14).

6.5 Let $\lambda = \lambda_i \mathbf{e}_i$ be a unit vector having (all positive) direction cosines λ_i with respect to the $\mathbf{e}_i$ coordinates. Define a new coordinate system $\mathbf{e}'_i$ so that the $\mathbf{e}'_3$ axis is in the direction λ. Define the $\mathbf{e}'_2$ axis so that it is perpendicular to $\mathbf{e}'_3$ and to $\mathbf{e}_1$ and has a positive projection on $\mathbf{e}_2$. Define the $\mathbf{e}'_1$ axis so that $\mathbf{e}'_1 \mathbf{e}'_2 \mathbf{e}'_3$ form a right-handed system (i.e., $\mathbf{e}'_1 \cdot \mathbf{e}'_2 \times \mathbf{e}'_3 = 1$). Show that the components of the rotation tensor $\mathbf{e}'_i = \mathbf{A} \cdot \mathbf{e}_i$ that transforms the $\mathbf{e}_i$ system to the $\mathbf{e}'_i$ system are

$$A(\lambda_1, \lambda_2, \lambda_3) = \begin{bmatrix} +\sqrt{\lambda_2^2 + \lambda_3^2} & 0 & \lambda_1 \\[2ex] -\dfrac{\lambda_1 \lambda_2}{\sqrt{\lambda_2^2 + \lambda_3^2}} & +\dfrac{\lambda_3}{\sqrt{\lambda_2^2 + \lambda_3^2}} & \lambda_2 \\[2ex] -\dfrac{\lambda_1 \lambda_3}{\sqrt{\lambda_2^2 + \lambda_3^2}} & -\dfrac{\lambda_2}{\sqrt{\lambda_2^2 + \lambda_3^2}} & \lambda_3 \end{bmatrix}$$

6.6 Given that the u_i are components of vectors (and, hence, transform like vector components when the coordinate system is rotated), show that the three quantities v_k defined by

$$u_k v_k = \alpha$$

where α is a scalar, are also components of a vector.

6.7 If F_{ij} are rectangular Cartesian components of a second-order tensor, show that if the nine quantities G_{ij} satisfy

$$F_{pq}G_{qp} = \alpha$$

where α is a scalar, then the G_{ij} are also components of a second-order tensor.

6.8 If the u_i and v_i are components of vectors (and, hence, transform like vector components when the coordinate system is rotated), show that the three quantities w_k defined by

$$w_k = \epsilon_{ijk}u_iv_j$$

are also components of a vector.

6.9 **F** and **G** are second-order tensors. If the components of **F** and **G** are related by

$$F_{ij} = H_{ijkl}G_{kl}$$

derive the rule for transforming the components H_{ijkl} from the unprimed to the primed system.

6.10 Show that if the components of a tensor **F** are symmetric in one rectangular Cartesian coordinate system, i.e., $F_{ij} = F_{ji}$, then its components are symmetric in *any* rectangular Cartesian system, i.e., $F'_{ij} = F'_{ji}$.

6.11 Prove that a tensor with components F_{ij} that are antisymmetric in one Cartesian coordinate system, i.e., $F_{ji} = -F_{ij}$, are antisymmetric in any Cartesian coordinate system.

6.12 Show that the sum of the normal components of a tensor is independent of the coordinate system used to express the components; that is,

$$F_{11} + F_{22} + F_{33} = F'_{11} + F'_{22} + F'_{33}$$

where F_{ij} and F'_{ij} are the components of **F** in any two rectangular Cartesian coordinate systems.

6.13 If F_{ij} are components of a second-order tensor, show that the combination $F_{kl}F_{lk}$ has the same value in any rectangular Cartesian coordinate system.

6.14 Consider a rotation of axis through an angle θ about the x_3 axis as shown in Figure 6.2. Show that the dyadic form of the tensor **A** in the relation

$$e'_k = \mathbf{A} \cdot \mathbf{e}_k$$

is given by $\mathbf{A} = \cos\theta(\mathbf{e}_1\mathbf{e}_1 + \mathbf{e}_2\mathbf{e}_2) + \sin\theta(\mathbf{e}_2\mathbf{e}_1 - \mathbf{e}_1\mathbf{e}_2) + \mathbf{e}_3\mathbf{e}_3$.

6.15 For the coordinate transformation in the preceding question:
(a) Determine the components of a vector **u** in the primed system having components u_i in the unprimed system.

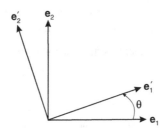

Figure 6.2   Rotation of the base vectors $\mathbf{e}_i$ to a new system $\mathbf{e}_i'$.

(b) Determine the components of a tensor $\mathbf{F}$ in the primed system having components F_{ij} in the unprimed system.

(c) Simplify your results in (a) and (b) for the case in which the rotation is small, $\theta = \mathrm{d}\theta \ll 1$, so that $\sin \mathrm{d}\theta \approx \mathrm{d}\theta$ and $\cos \mathrm{d}\theta \approx 1$.

6.16 The derivative of the orthogonal tensor $\mathbf{A}$ in Problem 6.14 with respect to θ is obtained by differentiating each element separately. If this derivative is $\dot{\mathbf{A}}(\theta)$ show that $\dot{\mathbf{A}}(\theta) = \mathbf{A}(\theta)\dot{\mathbf{A}}(0)$.

6.17 If the primed orthonormal base vectors are functions of time, $\mathbf{e}_p'(t) = \mathbf{A}(t) \cdot \mathbf{e}_p$, show that

$$\frac{\mathrm{d}}{\mathrm{d}t}\mathbf{e}_p'(t) = \dot{\mathbf{A}} \cdot \mathbf{A}^T \cdot \mathbf{e}_p'(t)$$

and that $\dot{\mathbf{A}} \cdot \mathbf{A}^T$ is skew symmetric.

6.18 **(a)** Determine the components of the orthogonal tensor corresponding to a $90°$ rotation about the x_2 axis.

(b) Determine the components of the orthogonal tensor corresponding to a $90°$ rotation about the x_3 axis.

(c) Determine the components of the orthogonal tensor corresponding to successive rotations: first $90°$ about the x_2 axis, then $90°$ about the x_3 axis.

6.19 Problem 6.5 derived the matrix transforming the unprimed coordinate system into one in which the x_3' axis was in the direction of a unit vector $\lambda = \lambda_i \mathbf{e}_i$.

(a) Combine this result with that of Problem 6.14 to show that the orthogonal tensor corresponding to causing rotation of the unprimed coordinate system through an angle θ about λ has components given by

$$\Lambda_{ij} = \delta_{ij} \cos \theta + \lambda_i \lambda_j (1 - \cos \theta) - \epsilon_{ijk} \lambda_k \sin \theta$$

(b) Show that tensor can be expressed in coordinate-free form as

$$\mathbf{\Lambda} = \mathbf{e}_3'\mathbf{e}_3' + (\mathbf{e}_1'\mathbf{e}_1' + \mathbf{e}_2'\mathbf{e}_2') \cos \theta - (\mathbf{e}_1'\mathbf{e}_2' - \mathbf{e}_2'\mathbf{e}_1') \sin \theta$$

where $\lambda = \mathbf{e}_3'$. This is a general representation of an orthogonal tensor and one can verify that it has the appropriate properties.

6.20 Show that two successive rotations of Problem 6.18 are equivalent to a rotation of 120° about a line making equal angles with the coordinate axis.

6.21 Show that αe_{ijk}, where α is a scalar, is an isotropic tensor of third order.

Reference

Aris R 1989 *Vectors, Tensors, and the Basic Equations of Fluid Mechanics*. Dover.

7

Principal Values and Principal Directions

In Chapter 2 we defined a tensor in terms of a black box (Figure 2.1) that takes a vector as input and produces a vector as output by means of certain rules. In general, the output vector has neither the same direction nor the same magnitude as the input vector. If, however, the output vector has the same direction, then, as discussed in Chapter 2, $\mathbf{v} = \lambda \mathbf{u}$, where λ is a scalar and

$$(\mathbf{F} - \lambda \mathbf{I}) \cdot \boldsymbol{\mu} = \mathbf{0} \tag{7.1}$$

where $\boldsymbol{\mu}$ can be taken as a unit vector.

The principal values (eigenvalues) λ_K and principal directions (eigenvectors) $\boldsymbol{\mu}_K$ of a second-order tensor $\mathbf{F}$ satisfy (2.15)

$$\mathbf{F} \cdot \boldsymbol{\mu}_K = \lambda_K \boldsymbol{\mu}_K \text{ (no sum on } K) \tag{7.2}$$

for $K = I, II, III$ and $\boldsymbol{\mu}_K$ can be taken as a unit vector without loss of generality.

Forming the scalar product of (7.2) with $\boldsymbol{\mu}_K$ yields

$$\boldsymbol{\mu}_K \cdot \mathbf{F} \cdot \boldsymbol{\mu}_K = \lambda_K \text{ (no sum on } K) \tag{7.3}$$

and with $\boldsymbol{\mu}_L \neq \boldsymbol{\mu}_K$ yields

$$\boldsymbol{\mu}_L \cdot \mathbf{F} \cdot \boldsymbol{\mu}_K = \lambda_K (\boldsymbol{\mu}_L \cdot \boldsymbol{\mu}_K) = 0 \text{ (no sum on } K) \tag{7.4}$$

Comparison to (3.10) identifies (7.3) and (7.4) as the components of $\mathbf{F}$ in a coordinate system aligned with the principal axes. Because the principal directions are orthonormal, they can be used as unit base vectors and $\mathbf{F}$ has the dyadic representation

$$\mathbf{F} = \lambda_I \boldsymbol{\mu}_I \boldsymbol{\mu}_I + \lambda_{II} \boldsymbol{\mu}_{II} \boldsymbol{\mu}_{II} + \lambda_{III} \boldsymbol{\mu}_{III} \boldsymbol{\mu}_{III} \tag{7.5}$$

Fundamentals of Continuum Mechanics, First Edition. John W. Rudnicki.
© 2015 John Wiley & Sons, Ltd. Published 2015 by John Wiley & Sons, Ltd.

Expressed differently, the matrix of components in this system is diagonal:

$$[F] = \begin{bmatrix} \lambda_I & 0 & 0 \\ 0 & \lambda_{II} & 0 \\ 0 & 0 & \lambda_{III} \end{bmatrix}$$

The dyadic representation of the orthogonal tensor that rotates the original basis system $\mathbf{e}_k$ into one aligned with the principal directions is

$$\mathbf{A} = \mu_K \mathbf{e}_k \tag{7.6}$$

where the k are still summed even though one is upper case and one lower. If the μ_K are expressed in terms of the $\mathbf{e}_i$ base vectors, then (7.6) becomes

$$\mathbf{A} = \left(\mu_K\right)_i \mathbf{e}_i \mathbf{e}_k$$

where $\left(\mu_k\right)_i$ is the ith component of the Kth eigenvector (relative to the $\mathbf{e}_i$ basis). Therefore the components of μ_K are the columns of $\mathbf{A}$ when written as a matrix. Expressing μ_K in terms of the $\mathbf{e}_i$ in (7.3) gives

$$\lambda_K = \left(\mu_K\right)_i F_{ij} \left(\mu_K\right)_j \text{ (no sum on } K)$$

In other words, using (7.6) as a coordinate transformation yields a diagonal form for the components of $\mathbf{F}$.

A non-trivial solution for (7.2) is possible only if the inverse of $\mathbf{F} - \lambda\mathbf{I}$ does not exist. As noted following (5.8), this requires that the determinant formed from the matrix of components vanishes:

$$\det\left(F_{ij} - \lambda\delta_{ij}\right) = 0 \tag{7.7}$$

Using the result of Problem 5.8 to expand (7.7) yields

$$\lambda^3 - I_1\lambda^2 - I_2\lambda - I_3 = 0 \tag{7.8}$$

where the coefficients are

$$I_1 = \operatorname{tr}\mathbf{F} = F_{kk} = F_{11} + F_{22} + F_{33} \tag{7.9}$$

$$I_2 = \frac{1}{2}(F_{ij}F_{ji} - F_{ii}F_{jj}) = \frac{1}{2}\left(\mathbf{F}\cdot\cdot\mathbf{F} - I_1^2\right) \tag{7.10}$$

$$I_3 = \det(\mathbf{F}) = \frac{1}{6}\epsilon_{ijk}\epsilon_{pqr}F_{ip}F_{jq}F_{kr} \tag{7.11}$$

Because the principal values are independent of the coordinate system, so are the coefficients in the characteristic equation (7.8) used to determine them. These coefficients are scalar invariants of the tensor $\mathbf{F}$ (generally called the *principal invariants*, since any combination of them is also invariant).

Using the principal axis representation of $\mathbf{F}$ (7.5) to form the tensor product of $\mathbf{F}$ with itself gives

$$\mathbf{F} \cdot \mathbf{F} = \left(\lambda_I\right)^2 \boldsymbol{\mu}_I \boldsymbol{\mu}_I + \left(\lambda_{II}\right)^2 \boldsymbol{\mu}_{II} \boldsymbol{\mu}_{II} + \left(\lambda_{III}\right)^2 \boldsymbol{\mu}_{III} \boldsymbol{\mu}_{III}$$

and the triple product is

$$\mathbf{F} \cdot \mathbf{F} \cdot \mathbf{F} = \left(\lambda_I\right)^3 \boldsymbol{\mu}_I \boldsymbol{\mu}_I + \left(\lambda_{II}\right)^3 \boldsymbol{\mu}_{II} \boldsymbol{\mu}_{II} + \left(\lambda_{III}\right)^3 \boldsymbol{\mu}_{III} \boldsymbol{\mu}_{III} \qquad (7.12)$$

Because each of the principal values satisfies (7.8), rearranging (7.12) for each of the principal values gives

$$\mathbf{F} \cdot \mathbf{F} \cdot \mathbf{F} = I_1 \mathbf{F} \cdot \mathbf{F} + I_2 \mathbf{F} + I_3 \mathbf{I} \qquad (7.13)$$

This is the *Cayley–Hamilton theorem*. A consequence is that $\mathbf{F}^N$, where $N > 3$, can be written as a sum of $\mathbf{F} \cdot \mathbf{F}$, $\mathbf{F}$, and $\mathbf{I}$ with coefficients that are functions of the invariants.

7.1   Example

Determine the principal values and principal directions for the tensor with the values shown in the matrix below:

$$[F] = \begin{bmatrix} 7 & 0 & -2 \\ 0 & 5 & 0 \\ -2 & 0 & 4 \end{bmatrix}$$

Expanding about the second row or column to take the determinant $\det(\mathbf{F} - \lambda\mathbf{I})$ gives

$$(5 - \lambda)\left[(7 - \lambda)(4 - \lambda) - (-2)(-2)\right] = 0$$

Therefore, the roots are $\lambda_I = 8$, $\lambda_{II} = 5$, and $\lambda_{III} = 3$. To determine the principal directions we substitute the principal values back in (7.1). For $\lambda_I = 8$ this gives the three equations

$$(7 - 8)\left(\mu^I\right)_1 + 0\left(\mu^I\right)_2 - 2\left(\mu^I\right)_3 = 0$$
$$0 + (5 - 8)\left(\mu^I\right)_2 - 0 = 0$$
$$-2\left(\mu^I\right)_1 + 0 + (4 - 8)\left(\mu^I\right)_3 = 0$$

Because the third equation is two times the first, these are linearly dependent. Either the first or third equation gives $\left(\boldsymbol{\mu}^{I}\right)_{1} = -2\left(\boldsymbol{\mu}^{I}\right)_{3}$ and the second gives $\left(\boldsymbol{\mu}^{I}\right)_{2} = 0$. Making $\boldsymbol{\mu}^{I}$ a unit vector gives

$$\boldsymbol{\mu}^{I} = \mp \frac{2}{\sqrt{5}}\mathbf{e}_{1} + 0\mathbf{e}_{2} \pm \frac{1}{\sqrt{5}}\mathbf{e}_{3}$$

For $\lambda_{II} = 5$, $\boldsymbol{\mu}^{II} = \pm\mathbf{e}_{2}$. This result can be recognized immediately since the off-diagonal elements of the second row and column are zero. We can follow the same procedure for $\boldsymbol{\mu}^{III}$, but it is simplest to choose $\boldsymbol{\mu}^{III} = \boldsymbol{\mu}^{I} \times \boldsymbol{\mu}^{II}$ for a right-handed system:

$$\boldsymbol{\mu}^{III} = \mp\frac{1}{\sqrt{5}}\mathbf{e}_{1} \mp \frac{2}{\sqrt{5}}\mathbf{e}_{3}$$

F is given in principal axis form as

$$\mathbf{F} = \lambda_{I}\boldsymbol{\mu}^{I}\boldsymbol{\mu}^{I} + \lambda_{II}\boldsymbol{\mu}^{II}\boldsymbol{\mu}^{II} + \lambda_{III}\boldsymbol{\mu}^{III}\boldsymbol{\mu}^{III}$$

The matrix with the components of the eigenvectors as columns is given by

$$[A] = \begin{bmatrix} -2/\sqrt{5} & 0 & -1/\sqrt{5} \\ 0 & 1 & 0 \\ 1/\sqrt{5} & 0 & -2/\sqrt{5} \end{bmatrix}$$

where we have taken the top sign. Taking the bottom signs yields the same result. Matrix multiplication can be used to verify that

$$\begin{bmatrix} -2/\sqrt{5} & 0 & 1/\sqrt{5} \\ 0 & 1 & 0 \\ -1/\sqrt{5} & 0 & -2/\sqrt{5} \end{bmatrix}\begin{bmatrix} 7 & 0 & -2 \\ 0 & 5 & 0 \\ -2 & 0 & 4 \end{bmatrix}\begin{bmatrix} -2/\sqrt{5} & 0 & -1/\sqrt{5} \\ 0 & 1 & 0 \\ 1/\sqrt{5} & 0 & -2/\sqrt{5} \end{bmatrix}$$

$$= \begin{bmatrix} 8 & 0 & 0 \\ 0 & 5 & 0 \\ 0 & 0 & 3 \end{bmatrix}$$

Exercises

7.1 In the example above, prove that $\left(\boldsymbol{\mu}^{II}\right)_{1} = \left(\boldsymbol{\mu}^{II}\right)_{3} = 0$.

7.2 Prove that the invariants are independent of coordinate system from (6.11). That is, show that the combination of components that forms an invariant is the same in the primed and unprimed systems.

7.3 Use the Cayley–Hamilton theorem (7.13) to obtain the following useful expression for the determinant:

$$\det (\mathbf{F}) = \frac{1}{3} \left\{ \text{tr} \, (\mathbf{F} \cdot \mathbf{F} \cdot \mathbf{F}) - 3 I_1 I_2 - I_1^3 \right\}$$

7.4 Use the Cayley–Hamilton theorem (7.13) to derive the following expressions:

(a) $\mathbf{F} \cdot \mathbf{F} = I_1 \mathbf{F} + I_2 \mathbf{I} + I_3 \mathbf{F}^{-1}$

(b) $\text{tr} \left(\mathbf{F}^{-1} \right) = -I_2 / I_3$

7.5 Show that if a tensor $\mathbf{F}$ has principal values λ_I, $\lambda_{II} = \lambda_{III}$, then the principal directions corresponding to λ_{II} and λ_{III} can be any vectors orthogonal to the principal direction of λ_I.

7.6

(a) Determine the principal values and principal directions of the tensor $\mathbf{F}$ with Cartesian components given by the matrix

$$[F] = \begin{bmatrix} 11 & 4 & 0 \\ 4 & 5 & 0 \\ 0 & 0 & 7 \end{bmatrix}$$

(b) Show that using the matrix with the components of the eigenvectors as columns as a rotation gives a diagonal form for $[F]$.

7.7 A tensor is given by the dyad $\mathbf{F} = \mathbf{ab}$. Determine the trace, determinant, and second invariant of $\mathbf{F}$.

7.8 Show that the only *real* principal value of the tensor of Problem 6.14 is one and that the corresponding principal direction is $\mathbf{e}_3$.

7.9 Use the determinant expansion of Problem 5.8 to show that the coefficients in (7.8) are given by (7.9), (7.10), and (7.11).

8

Gradient

Typically, the vectors and tensors used in continuum mechanics will be functions of position. Consequently, it is necessary to define an operation that expresses their changes with position. To do so, we first consider a scalar-valued function

$$\phi(\mathbf{x}) = \phi(x_1, x_2, x_3)$$

Figure 8.1 is a schematic of three level surfaces; that is, three surfaces on which the value of $\phi(\mathbf{x})$ is constant. A concrete example is a topographic map: ϕ is elevation and a function of two spatial variables. Contours indicate the positions of constant elevation. Now consider the change in ϕ as the position is changed from $\mathbf{x}$ to $\mathbf{x} + d\mathbf{x}$: write $d\mathbf{x} = \boldsymbol{\mu} \, ds$ where $\boldsymbol{\mu}$ is a unit vector in the direction of $d\mathbf{x}$ and ds is the magnitude of $d\mathbf{x}$. The change in ϕ per unit distance is defined by

$$\frac{d\phi}{ds} = \lim_{ds \to 0} \frac{\phi(\mathbf{x} + \boldsymbol{\mu} \, ds) - \phi(\mathbf{x})}{ds}$$

Writing

$$\frac{d\phi}{ds} = \boldsymbol{\mu} \cdot \nabla \phi \qquad (8.1)$$

defines the gradient of ϕ as $\nabla \phi$. The representation (8.1) is coordinate free. To convert to a Cartesian representation we expand the left side as

$$\frac{d\phi}{ds} = \left(\frac{\partial \phi}{\partial x_k} \mathbf{e}_k \right) \cdot \left(\frac{dx_l}{ds} \mathbf{e}_l \right) \qquad (8.2)$$

Noting that the second term in (8.2) is the Cartesian representation of $\boldsymbol{\mu}$ identifies the Cartesian component form of the gradient of ϕ as

$$\nabla \phi = \mathbf{e}_l \frac{\partial \phi}{\partial x_l} = \mathbf{e}_l \phi_{,l}$$

Fundamentals of Continuum Mechanics, First Edition. John W. Rudnicki.
© 2015 John Wiley & Sons, Ltd. Published 2015 by John Wiley & Sons, Ltd.

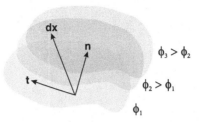

Figure 8.1 Schematic showing three level surfaces of the function ϕ. The normal **n** and tangent **t** are also shown with the infinitesimal change of position vector d**x**.

where $\phi,_l \equiv \partial\phi/\partial x_l$. We can generalize and define a *gradient operator* as

$$\nabla \equiv \mathbf{e}_k \frac{\partial}{\partial x_k}$$

If $\boldsymbol{\mu}$ is any vector tangent to the level surface then there is no change in ϕ and

$$\frac{d\phi}{dt} = 0$$

where dt is an infinitesimal distance in the tangent direction. Hence $\nabla\phi$ must be in the direction **n** perpendicular to the level surface:

$$\nabla\phi = \alpha\mathbf{n} \tag{8.3}$$

Taking $\boldsymbol{\mu} = \mathbf{n}$ in (8.1) and using (8.3) yields

$$\frac{d\phi}{dn} = \mathbf{n}\cdot(\nabla\phi) = \mathbf{n}\cdot(\alpha\mathbf{n}) = \alpha$$

Therefore, $\nabla\phi$ is in the direction **n** (normal to a surface of constant ϕ) and has the magnitude dϕ/dn:

$$\nabla\phi = \frac{d\phi}{dn}\mathbf{n}$$

This property makes the gradient useful for determining the normal to surfaces of constant ϕ.

An expression for the result of applying the gradient operator to a vector **v** follows naturally from the representation of tensors as dyadics:

$$\nabla\mathbf{v} = \left(\mathbf{e}_k\frac{\partial}{\partial x_k}\right)(v_l\mathbf{e}_l) = \frac{\partial v_l}{\partial x_k}\mathbf{e}_k\mathbf{e}_l = \partial_k v_l\mathbf{e}_k\mathbf{e}_l \tag{8.4}$$

The second equality follows because the base vectors have fixed magnitude (unit vectors) and direction. The last equality introduces the notation $\partial_k(\ldots) \equiv \partial(\ldots)/\partial x_k$. Using either this notation or $(\ldots)_{,k}$ is useful for keeping the subscripts in the same order as the dyadic base vectors. The tensor (8.4) has Cartesian components in matrix form given by

$$[\nabla \mathbf{v}] = \begin{bmatrix} \partial v_1/\partial x_1 & \partial v_2/\partial x_1 & \partial v_3/\partial x_1 \\ \partial v_1/\partial x_2 & \partial v_2/\partial x_2 & \partial v_3/\partial x_2 \\ \partial v_1/\partial x_3 & \partial v_2/\partial x_3 & \partial v_3/\partial x_3 \end{bmatrix}$$

To motivate this representation and demonstrate that the result is, in fact, a tensor, consider the Taylor expansion of vector components about x_j^o

$$v_i(x_j) = v_i(x_j^o) + \frac{\partial v_i}{\partial x_k}(x_j^o)(x_k - x_k^o) + \ldots \tag{8.5}$$

or, in vector form,

$$\mathbf{v}(\mathbf{x}) = \mathbf{v}(\mathbf{x}^o) + (\mathbf{x} - \mathbf{x}^o) \cdot \nabla \mathbf{v}(\mathbf{x})^o + \ldots \tag{8.6}$$

(Note that the order of the subscripts in (8.5) dictates the position of $(\mathbf{x} - \mathbf{x}^o)$ in (8.6).) Because $\nabla \mathbf{v}$ associates a vector $\mathbf{v}(\mathbf{x}) - \mathbf{v}(\mathbf{x}_o)$ with a vector $\mathbf{x} - \mathbf{x}_o$ by means of a relation that is linear and homogeneous, it is a tensor. The transpose of this tensor is

$$(\nabla \mathbf{v})^T = \frac{\partial v_i}{\partial x_j}\mathbf{e}_i\mathbf{e}_j = v_{i,j}\mathbf{e}_i\mathbf{e}_j \tag{8.7}$$

The scalar product of $\nabla \mathbf{v}$ and $\mathbf{I}$ yields the divergence of the vector $\mathbf{v}$

$$\nabla \mathbf{v} \cdot \cdot \mathbf{I} = \left(\frac{\partial v_i}{\partial x_j}\mathbf{e}_j\mathbf{e}_i\right) \cdot \cdot (\delta_{kl}\mathbf{e}_k\mathbf{e}_l) = \frac{\partial v_k}{\partial x_k}$$

We can also form the divergence from the scalar product of ∇ and $\mathbf{v}$

$$\nabla \cdot \mathbf{v} = \frac{\partial v_k}{\partial x_k}$$

If the vector $\mathbf{v}$ is the gradient of a scalar function ϕ, i.e., $\mathbf{v} = \nabla \phi$, then

$$\nabla \cdot \nabla \phi = \left(\mathbf{e}_h \frac{\partial}{\partial x_h}\right) \cdot \left(\mathbf{e}_l \frac{\partial \phi}{\partial x_l}\right) = \frac{\partial^2 \phi}{\partial x_k \partial x_k} = \nabla^2 \phi$$

gives the Laplacian of ϕ. Forming the cross product of ∇ and $\mathbf{v}$ yields the curl of $\mathbf{v}$

$$\nabla \times \mathbf{v} = \frac{\partial v_j}{\partial x_i}\epsilon_{ijk}\mathbf{e}_k = \mathbf{e}_i \partial_j v_k \epsilon_{ijk}$$

Similar arguments can be used to interpret the gradient of a tensor. A Taylor expansion of the tensor $\mathbf{F}$ about $\mathbf{x}_o$ yields

$$F_{ij}(x_k) = F_{ij}\left(x_k^o\right) + \frac{\partial F_{ij}}{\partial x_l}\left(x_k^o\right)\left(x_l - x_l^o\right) + \ldots$$

and identifies

$$\nabla \mathbf{F} = (\mathbf{e}_i \partial_i)(F_{jk}\mathbf{e}_j\mathbf{e}_k) = \frac{\partial F_{jk}}{\partial x_i}\mathbf{e}_i\mathbf{e}_j\mathbf{e}_k$$

as a third-order tensor. Forming the scalar and vector products of ∇ with $\mathbf{F}$ yields

$$\nabla \cdot \mathbf{F} = \mathbf{e}_k \partial_k \cdot (F_{lm}\mathbf{e}_l\mathbf{e}_m) = \delta_{kl}\frac{\partial F_{lm}}{\partial x_k}\mathbf{e}_m = \frac{\partial F_{km}}{\partial x_k}\mathbf{e}_m$$

$$\nabla \times \mathbf{F} = \mathbf{e}_k \partial_k \times (F_{lm}\mathbf{e}_l\mathbf{e}_m) = \frac{\partial F_{lm}}{\partial x_k}\epsilon_{kln}\mathbf{e}_n\mathbf{e}_m$$

8.1   Example: Cylindrical Coordinates

For reasons mentioned earlier, we will almost exclusively consider rectangular Cartesian systems in which the lengths and orientations of base vectors are fixed. Nevertheless, the treatment discussed here can be extended to more general coordinate systems. Although the formal procedure is straightforward the details of bookkeeping are more involved. As a simple example, consider cylindrical coordinates with unit orthogonal base vectors $\mathbf{e}_r$, $\mathbf{e}_\theta$, and $\mathbf{e}_z$ as shown in Figure 8.2. The gradient operator is given by

$$\nabla = \mathbf{e}_r\frac{\partial}{\partial r} + \mathbf{e}_\theta\frac{1}{r}\frac{\partial}{\partial \theta} + \mathbf{e}_z\frac{\partial}{\partial z}$$

The first and third terms are in the same form as in rectangular coordinates; the middle term requires $1/r$ in order to make the dimensions of each term be the reciprocal of length.

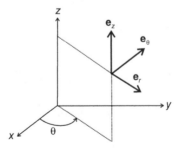

Figure 8.2   Base vectors in cylindrical coordinates depend on the angle θ.

In terms of base vectors in the x and y directions, the unit vectors in the r and θ directions are given by

$$\mathbf{e}_r = \cos\theta\,\mathbf{e}_x + \sin\theta\,\mathbf{e}_y, \quad \mathbf{e}_\theta = -\sin\theta\,\mathbf{e}_x + \cos\theta\,\mathbf{e}_y$$

Thus, the unit vectors $\mathbf{e}_\theta$ and $\mathbf{e}_r$ change with θ

$$\frac{d\mathbf{e}_r}{d\theta} = \mathbf{e}_\theta, \quad \frac{d\mathbf{e}_\theta}{d\theta} = -\mathbf{e}_r$$

Consequently, when applying the gradient operator to a vector in cylindrical coordinates, it is necessary to include the derivatives of the base vectors:

$$\nabla \cdot \mathbf{v} = \frac{\partial v_r}{\partial r} + \mathbf{e}_\theta \frac{1}{r}\frac{\partial}{\partial\theta} \cdot (v_r\mathbf{e}_r + v_\theta\mathbf{e}_\theta) + \frac{\partial v_z}{\partial z}$$

$$= \frac{\partial v_r}{\partial r} + \frac{v_r}{r} + \frac{1}{r}\frac{\partial v_\theta}{\partial\theta} + \frac{\partial v_z}{\partial z}$$

Similar operations can be used to generate the cylindrical coordinate forms for $\nabla \times \mathbf{v}$, $\nabla^2\phi$, $\nabla\mathbf{v}$, and operations of the gradient on tensors.

Exercises

8.1 Prove that if $\mathbf{v} = \nabla\phi$, then $\nabla \times \mathbf{v} = 0$.

8.2 If $\mathbf{F}$ is a symmetric second-order tensor, determine the unit normal to the surface given by

$$\mathbf{x} \cdot \mathbf{F} \cdot \mathbf{x} = 1$$

8.3 Show that the rectangular components of the vector $\mathbf{v} \cdot \nabla\mathbf{u}$, where the tensor $\nabla\mathbf{u}$ is the gradient of the vector $\mathbf{u}$, are the same as the result from applying to $\mathbf{u}$ the operator obtained formally as the product $\mathbf{v} \cdot \nabla$. In other words, show that

$$\mathbf{v} \cdot (\nabla\mathbf{u}) = (\mathbf{v} \cdot \nabla)\mathbf{u}$$

8.4 Use index notation to prove the following identities:
(a) $\nabla \cdot (\phi\mathbf{v}) = \mathbf{v} \cdot \nabla\phi + \phi\nabla \cdot \mathbf{v}$
(b) $\nabla \times (\phi\mathbf{v}) = \nabla\phi \times \mathbf{v} + \phi\nabla \times \mathbf{v}$

8.5 Use index notation to prove the following identities, where $\mathbf{u}$ and $\mathbf{v}$ are vectors and $\mathbf{F}$ is a second-order tensor:
(a) $\nabla \cdot (\mathbf{u} \times \mathbf{v}) = (\nabla \times \mathbf{u}) \cdot \mathbf{v} - \mathbf{u} \cdot (\nabla \times \mathbf{v})$
(b) $\nabla \cdot (\mathbf{F} \cdot \mathbf{u}) = \mathbf{u} \cdot (\nabla \cdot \mathbf{F}) + \mathbf{F}^T \cdot\cdot\nabla\mathbf{u}$

8.6 Use index notation to prove the following identities involving the gradient operator:
(a) $\nabla \times (\mathbf{u} \times \mathbf{v}) = \mathbf{u}(\nabla \cdot \mathbf{v}) - \mathbf{v}(\nabla \cdot \mathbf{u}) + (\mathbf{v} \cdot \nabla)\mathbf{u} - (\mathbf{u} \cdot \nabla)\mathbf{v}$
(b) $\nabla \times (\nabla \times \mathbf{v}) = \nabla(\nabla \cdot \mathbf{v}) - \nabla^2\mathbf{v}$ where $\nabla^2 = \nabla \cdot \nabla$

8.7 Use index notation to calculate
   (a) ∇r
   (b) $\nabla (\mathbf{x}/r)$
   (c) $\nabla^2 r = \nabla \cdot \nabla r$ where $r^2 = \mathbf{x} \cdot \mathbf{x} = x_k x_k$

8.8 Show that

$$\nabla \times \mathbf{u} = -\mathbf{u} \times \nabla$$

where the derivative in ∇ on the right side is to be interpreted as acting on $\mathbf{u}$ (but the order of the base vectors remains as written). Thus the usual property of the cross product applies if one of the multipliers is the gradient operator ∇.

8.9 Show that

$$\mathbf{F} \times \nabla = - \left(\nabla \times \mathbf{F}^T \right)^T$$

where $\mathbf{F}$ is a second-order tensor and interpretation of the ∇ acting from the right is the same as in Problem 8.8.

8.10 If $\mathbf{M} = \nabla \times \mathbf{E}$, write out the rectangular Cartesian components M_{11}, M_{22}, M_{12}, and M_{21}.

8.11 Determine the form of Laplacian $\nabla^2 \phi$ in cylindrical coordinates where ϕ is a scalar function.

8.12 Determine the form of curl $\nabla \times \mathbf{v}$ in cylindrical coordinates. Answer:

$$\mathbf{e}_r \left(\frac{1}{r} \frac{\partial v_z}{\partial \theta} - \frac{\partial v_\theta}{\partial z} \right) + \mathbf{e}_\theta \left(\frac{\partial v_r}{\partial z} - \frac{\partial v_z}{\partial r} \right) + \mathbf{e}_z \left(\frac{\partial v_\theta}{\partial r} - \frac{1}{r} \frac{\partial v_r}{\partial \theta} + \frac{v_\theta}{r} \right)$$

8.13 Determine the cylindrical component form of $\nabla \mathbf{v}$. Answer:

$$[\nabla \mathbf{v}] = \begin{bmatrix} \dfrac{\partial v_r}{\partial r} & \dfrac{\partial v_\theta}{\partial r} & \dfrac{\partial v_z}{\partial r} \\[2mm] \dfrac{1}{r} \dfrac{\partial v_r}{\partial \theta} - \dfrac{v_\theta}{r} & \dfrac{1}{r} \dfrac{\partial v_\theta}{\partial \theta} + \dfrac{v_r}{r} & \dfrac{1}{r} \dfrac{\partial v_z}{\partial \theta} \\[2mm] \dfrac{\partial v_r}{\partial z} & \dfrac{\partial v_\theta}{\partial z} & \dfrac{\partial v_z}{\partial z} \end{bmatrix}$$

8.14 Determine the form of the divergence of a (second-order) tensor $\nabla \cdot \mathbf{F}$ in cylindrical components. Answer:

$$\mathbf{e}_r \left\{ \frac{\partial F_{rr}}{\partial r} + \frac{F_{rr} - F_{\theta\theta}}{r} + \frac{1}{r} \frac{\partial F_{\theta r}}{\partial \theta} + \frac{\partial F_{zr}}{\partial z} \right\}$$

$$+ \mathbf{e}_\theta \left\{ \frac{\partial F_{r\theta}}{\partial r} + \frac{1}{r} \frac{\partial F_{\theta\theta}}{\partial \theta} + \frac{F_{r\theta}}{r} + \frac{F_{\theta r}}{r} + \frac{\partial F_{z\theta}}{\partial z} \right\} + \mathbf{e}_z \left\{ \frac{\partial F_{rz}}{\partial r} + \frac{1}{r} \frac{\partial F_{\theta z}}{\partial \theta} + \frac{F_{rz}}{r} + \frac{\partial F_{zz}}{\partial z} \right\}$$

Part Two

Stress

Tensors were introduced in Part One. This part defines and discusses stress as the first example of a particular physical tensor. Although stress is familiar from strength of materials, the emphasis here is on stress as a tensor. Because the stress discussed in this part is symmetric (later we will see that not all stress measures are symmetric), many of the results, in particular those in Chapters 10 and 11, apply to other symmetric tensors to be introduced later.

Fundamentals of Continuum Mechanics, First Edition. John W. Rudnicki.
© 2015 John Wiley & Sons, Ltd. Published 2015 by John Wiley & Sons, Ltd.

9

Traction and Stress Tensor

9.1 Types of Forces

We have already said that continuum mechanics assumes an actual body can be described by associating with it a mathematically continuous body. For example, we define the density at a point P as

$$\rho^{(P)} = \lim_{\Delta V \to 0} \frac{\Delta m}{\Delta V}$$

where ΔV contains the point P and Δm is the mass contained in ΔV. Continuum mechanics assumes that it makes sense, or at least is useful, to perform this limiting process even though we know that matter is discrete on an atomic scale and, often, on larger scales, e.g., granular materials or a fissured rock mass. More precisely, ρ is the average density in a representative volume around the point P. What is meant by a representative volume depends on the material being considered and the scale of interest. For example, we can model a polycrystalline material with a density that varies strongly from point to point in different grains. Alternatively, we might use a uniform density that reflects the average over several grains. In a fissured rock mass, the representative volume might be much larger, on the order of meters or tens of meters, encompassing a sufficient number of fissures.

Just as we have considered the mass to be distributed continuously, so also do we consider the forces to be continuously distributed. These may be of two types:

1. *Body forces* have a magnitude proportional to the mass and act at a distance, e.g., gravity, magnetic forces (Figure 9.1). Body forces are computed per unit mass **b** or per unit volume $\rho\mathbf{b}$:

$$\mathbf{b}(\mathbf{x}) = \lim_{\Delta m \to 0} \frac{\Delta \mathbf{f}}{\Delta m}$$

The continuum hypothesis asserts that this limit exists, has a unique value, and is independent of the manner in which $\Delta m \to 0$.

Fundamentals of Continuum Mechanics, First Edition. John W. Rudnicki.
© 2015 John Wiley & Sons, Ltd. Published 2015 by John Wiley & Sons, Ltd.

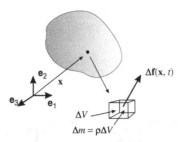

Figure 9.1   Illustration of the force $\Delta\mathbf{f}(\mathbf{x}, t)$ acting on the volume element ΔV.

2. *Surface forces* are computed per unit area and are contact forces. They may be forces that
   are applied to the exterior surface of the body or they may be forces transmitted from one
   part of a body to another.

Consider the forces acting on and within a body (Figure 9.2). We can define the traction by
means of the following conceptual procedure: Slice the body by a surface R (not necessarily
planar) that passes through the point Q and divides the body into parts 1 and 2. Remove part
1 and replace it by the forces that 1 exerts on 2. The forces that 2 exerts on 1 are equal and
opposite. Now consider the forces exerted by 1 on 2 on a portion of the surface having area
ΔS and normal $\mathbf{n}$ at Q. We can replace the distribution of forces on this surface by a statically
equivalent force $\Delta\mathbf{f}$ and moment $\Delta\mathbf{m}$ at Q. We define the *average traction* on ΔS as

$$\Delta\mathbf{t}^{(\text{avg})} = \frac{\Delta\mathbf{f}}{\Delta S}$$

Now we shrink C keeping point Q contained in C and define traction at a point Q by

$$\mathbf{t}^{(\mathbf{n})} = \lim_{\Delta S \to 0} \frac{\Delta\mathbf{f}}{\Delta S} \tag{9.1}$$

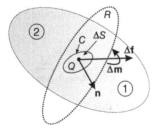

Figure 9.2   The surface R passes through the point Q and divides the body into two parts. The curve C
contains Q and encloses an area ΔS. The unit normal to the surface at Q is $\mathbf{n}$. The net force exerted by 1
on 2 across ΔS is $\Delta\mathbf{f}$ and the net moment is $\Delta\mathbf{m}$.

This is a vector that equals the force per unit area (intensity of force) exerted at Q by the material of 1 (side into which **n** points) on 2. The traction is often called the "stress vector" but we will use "stress" exclusively to refer to a tensor.

In taking the limit (9.1) we have assumed that it is independent of the manner in which $\Delta S \to 0$ and the choice of the surface ΔS as long as the normal at Q is unique. In addition we have assumed that $\Delta \mathbf{f}$ varies continuously, there is no concentrated force at Q (though this situation can be addressed), and that

$$\lim_{\Delta S \to 0} \frac{\Delta \mathbf{m}}{\Delta S} = 0$$

The last will necessarily be the case if the couple is due to distributed forces. The theory of couple stresses does not make this assumption (see, e.g., Malvern 1988, pp. 217–220).

9.2   Traction on Different Surfaces

The traction depends on the orientation of the normal through the point. To establish the result of reversing the normal, we will use Newton's second law

$$\sum \mathbf{F} = m \frac{d\mathbf{v}}{dt} \tag{9.2}$$

where $\mathbf{F}$ is the force, m is the mass, and $\mathbf{v}$ is the velocity. Now we apply this to a slice of material of thickness h and area ΔS (Figure 9.3):

$$\mathbf{t}^{(\mathbf{n})} \Delta S + \mathbf{t}^{(-\mathbf{n})} \Delta S + \rho \mathbf{b} \Delta S h = \rho \Delta S h \frac{d\mathbf{v}}{dt}$$

where we have written the mass as $\rho \Delta S h$. Dividing by ΔS yields

$$\mathbf{t}^{(\mathbf{n})} + \mathbf{t}^{(-\mathbf{n})} + \rho \mathbf{b} h = \rho h \frac{d\mathbf{v}}{dt}$$

Letting $h \to 0$ yields

$$\mathbf{t}^{(n)} = -\mathbf{t}^{(-n)} \tag{9.3}$$

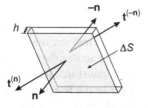

Figure 9.3   Tractions acting on opposite sides of a thin slice of material.

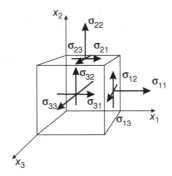

Figure 9.4   Illustration of the labeling of the components of the stress tensor. Remember that the cube shown represents a *point*.

Thus, the traction vectors are equal in magnitude and opposite in sign on two sides of a surface. In other words, reversing the direction of the normal to the surface reverses the sign of the traction vector. We can express the traction on planes normal to the coordinate directions $\mathbf{t}^{(\mathbf{e}_i)}$ in terms of their components (Figure 9.4)

$$\mathbf{t}^{(\mathbf{e}_1)} = \sigma_{11}\mathbf{e}_1 + \sigma_{12}\mathbf{e}_2 + \sigma_{13}\mathbf{e}_3$$
$$\mathbf{t}^{(\mathbf{e}_2)} = \sigma_{21}\mathbf{e}_1 + \sigma_{22}\mathbf{e}_2 + \sigma_{23}\mathbf{e}_3$$
$$\mathbf{t}^{(\mathbf{e}_3)} = \sigma_{31}\mathbf{e}_1 + \sigma_{32}\mathbf{e}_2 + \sigma_{33}\mathbf{e}_3$$

These three equations can be written more compactly using the summation convention as

$$\mathbf{t}^{(\mathbf{e}_i)} = \sigma_{ij}\mathbf{e}_j \tag{9.4}$$

where the first index i denotes the direction of the normal to the plane on which the force acts and the second index j denotes the direction of the force component. We can also express the traction as the scalar product of $\mathbf{e}_i$ with a tensor:

$$\mathbf{t}^{(i)} = \mathbf{e}_i \cdot (\sigma_{mn}\mathbf{e}_m\mathbf{e}_n)$$

The term in parentheses is the dyadic representation of the stress tensor σ and the σ_{ij} are its Cartesian components; $\sigma_{11}, \sigma_{22}, \sigma_{33}$ are *normal stresses*, and $\sigma_{12}, \sigma_{21}, \sigma_{32}, \sigma_{23}, \sigma_{31}, \sigma_{13}$ are *shear stresses*. Typically, in engineering, normal stresses are positive if they act in tension. In this case a stress component is positive if it acts in the positive coordinate direction on a face with outward normal in the positive coordinate direction, or if it acts in the negative coordinate direction on a face with outward normal in the negative coordinate direction. (Note that for a bar in equilibrium the forces and tractions acting on the ends of the bar are in opposite directions and have opposite signs, but these correspond to stress components of the same sign.) In geology or geotechnical engineering, the sign convention is often reversed because normal stresses are typically compressive.

9.3   Traction on an Arbitrary Plane (Cauchy Tetrahedron)

Equation (9.4) gives the tractions on planes with normals in the coordinate directions but we would like to determine the traction on a plane with a normal in an arbitrary direction. Figure 9.5 shows a tetrahedron with three faces perpendicular to the coordinate axes and the fourth (oblique) face with a normal vector **n**. The oblique face has area ΔS and the area of the other faces can be expressed as $\Delta S_i = n_i \Delta S$ (see Problem 1.7). The volume of the tetrahedron is $\Delta V = (1/3)h\Delta S$ where h is the distance perpendicular to the oblique face through the origin. Applying Newton's second Law (9.2) to this tetrahedron gives

$$\mathbf{t}^{(\mathbf{n})}\Delta S + (-\mathbf{t}^{(i)}\Delta S_i) + \rho\mathbf{b}\Delta V = \rho\Delta V \frac{d\mathbf{v}}{dt}$$

In the second term, we have used (9.3) to express the sum of the forces acting on the planes perpendicular to the negative of the coordinate directions. We divide by ΔS and let $h \to 0$. The result is

$$\mathbf{t}^{(\mathbf{n})} = \mathbf{t}^{(i)}n_i = n_1\mathbf{t}^{(1)} + n_2\mathbf{t}^{(2)} + n_3\mathbf{t}^{(3)}$$

Substituting (9.4) yields

$$\mathbf{t}^{(\mathbf{n})} = n_i\sigma_{ij}\mathbf{e}_j = \mathbf{n}\cdot\boldsymbol{\sigma}$$

This expression associates a vector $\mathbf{t}^{(\mathbf{n})}$ with every direction in space **n** by means of an expression that is linear and homogeneous and, hence, establishes $\boldsymbol{\sigma}$ as a tensor. Since the **n** appears on the right side we will omit it as a superscript on **t** hereafter. Because $\boldsymbol{\sigma}$ is a tensor,

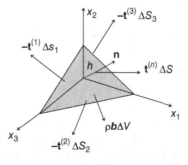

Figure 9.5   Tetrahedron with tractions acting on the faces.

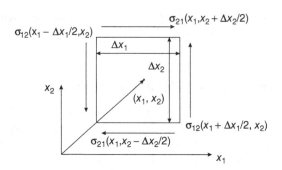

Figure 9.6 Shear stresses acting on a small element.

its components in a rectangular Cartesian system must transform according to (6.11) under a rotation of the coordinates:

$$\sigma'_{ij} = A_{pi}A_{qj}\sigma_{pq}$$

where

$$A_{pi} = \mathbf{e}'_i \cdot \mathbf{e}_p$$

9.4   Symmetry of the Stress Tensor

We can also show that σ is a symmetric tensor. To this end we require that the sum of the moments be equal to the moment of inertia multiplied by the angular acceleration for a small cuboidal element centered at (x_1, x_2, x_3) with edges Δx_1, Δx_2, and Δx_3 (Figure 9.6). For simplicity, consider the element to be subjected only to shear stresses σ_{12} and σ_{21} in the $x_1 x_2$ plane (Δx_3 is not shown). Taking account of the difference in position of the sides from the center and summing the moments yields

$$\left[\sigma_{12}\left(x_1 + \frac{\Delta x_1}{2}, x_2\right) + \sigma_{12}\left(x_1 - \frac{\Delta x_1}{2}, x_2\right)\right]\Delta x_2 \Delta x_3 \frac{1}{2}\Delta x_1$$

$$-\left[\sigma_{21}\left(x_1, x_2 + \frac{\Delta x_2}{2}\right) + \sigma_{21}\left(x_1, x_2 - \frac{\Delta x_2}{2}\right)\right]\Delta x_1 \Delta x_3 \frac{1}{2}\Delta x_2$$

$$= \alpha \frac{\rho}{12}(\Delta x_1 \Delta x_2 \Delta x_3)\left(\Delta x_1^2 + \Delta x_2^2\right)$$

where the $\Delta x_1/2$ and $\Delta x_2/2$ in the first two lines are the moment arms, and the third line is the angular acceleration α multiplied by the moment of inertia about the center. Dividing by $\Delta x_1 \Delta x_2 \Delta x_3$ and letting Δx_1, $\Delta x_2 \to 0$ yields $\sigma_{21} = \sigma_{12}$ and, similarly, $\sigma_{ij} = \sigma_{ji}$. Later we will give a more general derivation of this result and see that it does not pertain when the stress is defined per unit reference (as distinguished from current) area.

Exercise

9.1 The stress state in a body occupying $|x_1| \leq a$, $|x_2| \leq a$, $|x_3| \leq h$ is given by

$$\sigma_{11} = -p\frac{\left(x_1^2 - x_2^2\right)}{a^2}$$

$$\sigma_{22} = p\frac{\left(x_1^2 - x_2^2\right)}{a^2}$$

$$\sigma_{12} = 2px_1x_2/a^2$$

and $\sigma_{33} = \sigma_{13} = \sigma_{23} = 0$. Calculate the traction vector on the face $x_1 = +a$ and the face $x_1 = -a$.

Reference

Malvern LE 1988 *Introduction to the Mechanics of a Continuous Medium*. Prentice Hall.

10

Principal Values of Stress

Because σ is a symmetric tensor it has three real principal values with at least one set of orthogonal principal directions (see Chapter 2). The principal values λ and directions $\mathbf{n}$ satisfy (7.1), rewritten here in index form with the current notation:

$$(\sigma_{ij} - \lambda\delta_{ij})n_j = 0 \qquad (10.1)$$

Rearranging this equation shows that the directions $\mathbf{n}$ satisfying this equation are those for planes having only normal tractions. We will now rederive this equation by another approach. In doing so, we will show that two of the principal values correspond to the largest and smallest values of the normal stress on any plane. Because this derivation makes use of no special properties of the stress, the result applies to the principal values of any (real) symmetric tensor. Thus, it provides an alternative interpretation of the principal values as including the largest and smallest normal components of the tensor.

The normal component of $\mathbf{t}$ in the direction $\mathbf{n}$ is

$$t_n = \mathbf{n} \cdot \boldsymbol{\sigma} \cdot \mathbf{n} = n_i\sigma_{ij}n_j \qquad (10.2)$$

(Note that the subscript n is not an index here but designates the normal component.) We wish to find the largest and smallest values of t_n as the normal to the plane varies over all directions. The n_i are not, however, independent but are subject to the constraint

$$\mathbf{n} \cdot \mathbf{n} = n_i n_i = 1 \qquad (10.3)$$

We could deal with this problem by using (10.3) to eliminate one of the n_i or by using two angles with respect to a coordinate system. The drawback of these approaches is that they require specifying which of the n_i are to be eliminated or how to define the angles. A more elegant approach is to incorporate the constraint by using a Lagrange multiplier, here denoted σ. The constraint multiplied by σ is added to $n_i\sigma_{ij}n_j$. The derivative $\partial(\ldots)/\partial\sigma = 0$ then returns the constraint equation (10.3). Now the n_i can be treated as independent. The largest

Fundamentals of Continuum Mechanics, First Edition. John W. Rudnicki.
© 2015 John Wiley & Sons, Ltd. Published 2015 by John Wiley & Sons, Ltd.

and smallest (stationary) values of the normal traction (10.2) are obtained by differentiating with respect to the n_k:

$$\frac{\partial}{\partial n_k}\left\{n_i\sigma_{ij}n_j - \sigma(n_in_i - 1)\right\} = 0 \tag{10.4}$$

Carrying out the differentiation in (10.4) yields

$$\frac{\partial n_i}{\partial n_k}\sigma_{ij}n_j + n_i\sigma_{ij}\frac{\partial n_j}{\partial n_k} - 2\sigma n_i\frac{\partial n_i}{\partial n_k} = 0 \tag{10.5}$$

Recognizing that $\partial n_i/\partial n_k = \delta_{ik}$, we can rewrite (10.5) as

$$\sigma_{kj}n_j + n_i\sigma_{ik} - 2\sigma n_k = 0$$

or, after using the symmetry of σ_{ij},

$$(\sigma_{kj} - \sigma\delta_{kj})n_j = 0$$

which is the same as (10.1). Therefore, the principal values of σ_{ij}, the roots of

$$\det|\sigma_{kj} - \sigma\delta_{kj}| = 0$$

are the stationary values of t_n. We denote these roots by

$$\sigma_I > \sigma_{II} > \sigma_{III}$$

with corresponding principal directions $\mathbf{n}^{(I)}, \mathbf{n}^{(II)}$, and $\mathbf{n}^{(III)}$. σ_I (σ_{III}) is the largest (smallest) normal stress. σ_{II} is a stationary value, i.e., the largest normal stress in the plane defined by $\mathbf{n}^{(II)}$ and $\mathbf{n}^{(III)}$ and the smallest normal stress in the plane defined by $\mathbf{n}^{(I)}$ and $\mathbf{n}^{(II)}$. If two of the principal values are equal, say $\sigma_I = \sigma_{II}$, then the direction $\mathbf{n}^{(III)}$ is unique, but any rotation about $\mathbf{n}^{(III)}$ yields another set of principal axes.

From (7.8) we know that the principal values satisfy

$$\sigma^3 - I_1\sigma^2 - I_2\sigma - I_3 = 0 \tag{10.6}$$

where the coefficients are given by (7.9), (7.10), and (7.11) and other results from Chapter 7 also apply here.

10.1   Deviatoric Stress

It is often useful to separate the stress (or, indeed, any tensor) into a part with zero trace, called the *deviatoric* part, and an *isotropic* tensor (see Section 6.4). The deviatoric stress is defined as

$$\sigma'_{ij} = \sigma_{ij} - \frac{1}{3}\delta_{ij}\sigma_{kk}$$

or

$$\sigma' = \sigma - \frac{1}{3}(\text{tr }\sigma)\mathbf{I}$$

By construction, the trace of the deviator, the first invariant, vanishes

$$\text{tr }\sigma' = \mathbf{I} \cdot \cdot\sigma' = 0 \tag{10.7}$$

Unless the equation for the principal values (10.6) is easy to factor, it is generally more convenient to solve numerically. It is, however, possible to obtain a closed form solution for the principal values of the deviatoric stress. Because the first invariant of the deviatoric stress vanishes, i.e., (10.7), the equation for the principal values becomes

$$s^3 - J_2 s - J_3 = 0 \tag{10.8}$$

where s is the principal value and the invariants J_2 and J_3 are given by

$$J_2 = \frac{1}{2}\text{tr}(\sigma' \cdot \sigma') = \frac{1}{2}\sigma'_{ij}\sigma'_{ji} \tag{10.9}$$

$$J_3 = \det(\sigma'_{ij}) = \frac{1}{3}\text{tr}(\sigma' \cdot \sigma' \cdot \sigma') = \frac{1}{3}\sigma'_{ik}\sigma'_{kl}\sigma'_{li} \tag{10.10}$$

The first of these (10.9) follows from (7.10) and the second from Problem 7.3. Making the substitution

$$s = \left(\frac{4}{3}J_2\right)^{1/2}\sin\alpha \tag{10.11}$$

in (10.8) and using some trigonometric identities yields

$$\sin 3\alpha = \frac{-\sqrt{27J_3}}{2(J_2)^{3/2}} \tag{10.12}$$

or

$$\alpha = \frac{1}{3}\arcsin\left(-\frac{\sqrt{27J_3}}{2(J_2)^{3/2}}\right) \tag{10.13}$$

This yields one root of (10.8). Two additional roots are given by $\alpha \pm 2\pi/3$.

10.2   Example

Consider the stress tensor

$$\sigma = \frac{1}{2}\alpha\left(\lambda\mu + \mu\lambda\right)$$

where α is a scalar and λ and μ are unit vectors. Show that the principal direction of the greatest principal value bisects λ and μ.

To simplify the calculation, we can take $\lambda = e_1$ and e_3 perpendicular to the plane of λ and μ. The stress then takes the form

$$\sigma = \alpha \left\{ \cos(\theta) e_1 e_1 + \frac{1}{2} \sin(\theta)(e_1 e_2 + e_2 e_1) \right\}$$

where θ is the angle between λ and μ. The principal values are $\alpha \cos^2(\theta/2), 0,$ and $-\alpha \sin^2(\theta/2)$. The principal direction corresponding to $\cos^2(\theta/2)$ is

$$\cos(\theta/2) e_1 + \sin(\theta/2) e_2$$

and hence bisects λ and μ.

Exercises

10.1 The components of the stress tensor at a point are given by

$$[\sigma] = \begin{bmatrix} -1 & -2 & 0 \\ -2 & 0 & 2 \\ 0 & 2 & 1 \end{bmatrix}$$

(a) Determine the principal stresses.
(b) Determine the principal directions.

10.2 The stress tensor is given by

$$\sigma = \tau \cos \theta (e_1 e_2 + e_2 e_1) + \tau \sin \theta (e_1 e_3 + e_3 e_1)$$

(a) Determine the principal stresses.
(b) Determine the principal directions.

10.3 Show that J_2 can be written in terms of the principal stresses as

$$J_2 = \frac{1}{6} \left\{ (\sigma_1 - \sigma_2)^2 + (\sigma_2 - \sigma_3)^2 + (\sigma_3 - \sigma_1)^2 \right\}$$

10.4 Fill in the details of using (10.11) in (10.8) to obtain (10.12).

10.5 Determine the value of the angle α in (10.13) for the following stress states [Hint: Use the result of Problem 10.3]:
(a) $\sigma_2 = (1/2)(\sigma_1 + \sigma_3)$
(b) $\sigma_1 > \sigma_2 = \sigma_3$
(c) $\sigma_1 = \sigma_2 > \sigma_3$

10.6 Show that the principal directions for the deviatoric stress are the same as those for the stress.

11

Stationary Values of Shear Traction

In Chapter 10, we answered the question "For a given state of stress what are the orientations of the planes that have the maximum and minimum normal stress?" We found that these are the principal planes, planes on which there is no shear traction. We can ask a similar question about the shear traction: "What is the maximum value of the shear traction and on what plane does it occur?"

The traction on a plane with a normal $\mathbf{n}$ can be resolved into a normal component $t_n = \mathbf{n} \cdot \mathbf{t}$, where $\mathbf{t} = \mathbf{n} \cdot \boldsymbol{\sigma}$, and shear a component t_s

$$\mathbf{t} = (\mathbf{n} \cdot \mathbf{t})\mathbf{n} + t_s \mathbf{s}$$

where $\mathbf{n} \cdot \mathbf{s} = 0$ (Figure 11.1). Rearranging as

$$t_s \mathbf{s} = \mathbf{t} - t_n \mathbf{n}$$

and forming the scalar product of each side with itself yields the square of the magnitude of the shear traction

$$t_s^2 = \mathbf{t} \cdot \mathbf{t} - t_n^2$$

or, in component form in terms of the stress,

$$t_s^2 = (n_p \sigma_{pq})(n_r \sigma_{rq}) - (n_p \sigma_{pq} n_q)^2 \tag{11.1}$$

Because the sign of the shear traction has no physical significance (unlike the sign of the normal traction, which indicates tension or compression), there is no loss of generality in working with the square of the shear traction.

Just as we did for the normal stress, we want to let $\mathbf{n}$ vary over all directions and find the largest and smallest values of the shear traction. Thus, we want to find stationary values of the shear traction, subject to the condition

$$\mathbf{n} \cdot \mathbf{n} = n_k n_k = 1$$

Fundamentals of Continuum Mechanics, First Edition. John W. Rudnicki.
© 2015 John Wiley & Sons, Ltd. Published 2015 by John Wiley & Sons, Ltd.

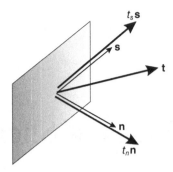

Figure 11.1   Traction on a plane with normal **n** resolved into shear and normal components.

To facilitate the calculation, we choose the principal directions as coordinate axes $\mathbf{e}_1$, $\mathbf{e}_2$, and $\mathbf{e}_3$ with corresponding principal stresses σ_1, σ_2, and σ_3. Consequently, the stress tensor and unit normal to the plane can be written as

$$\boldsymbol{\sigma} = \sum_k \sigma_k \mathbf{e}_k \mathbf{e}_k = \sigma_1 \mathbf{e}_1 \mathbf{e}_1 + \sigma_2 \mathbf{e}_2 \mathbf{e}_2 + \sigma_3 \mathbf{e}_3 \mathbf{e}_3 \tag{11.2}$$

$$\mathbf{n} = \sum_k n_k \mathbf{e}_k = n_1 \mathbf{e}_1 + n_2 \mathbf{e}_2 + n_3 \mathbf{e}_3 \tag{11.3}$$

Summation is indicated explicitly in (11.2) since the subscript "k" appears three times and, for clarity, summation is also explicit in (11.3). Then the traction on the plane with normal **n** is

$$\mathbf{t} = \mathbf{n} \cdot \boldsymbol{\sigma} = \sum_l n_l \mathbf{e}_l \cdot \sum_k \sigma_k \mathbf{e}_k \mathbf{e}_k$$

$$= \sum_k n_k \sigma_k \mathbf{e}_k = n_1 \sigma_1 \mathbf{e}_1 + n_2 \sigma_2 \mathbf{e}_2 + n_3 \sigma_3 \mathbf{e}_3$$

The normal traction is

$$t_n = \mathbf{n} \cdot \boldsymbol{\sigma} \cdot \mathbf{n} = \sum_k n_k^2 \sigma_k = n_1^2 \sigma_1 + n_2^2 \sigma_2 + n_3^2 \sigma_3$$

The first term of (11.1) is

$$\sum_k n_k \sigma_k \mathbf{e}_k \cdot \sum_k n_k \sigma_k \mathbf{e}_k = \sum_k n_k^2 \sigma_k{}^2$$

Forming the shear traction (11.1) and taking the derivative

$$\frac{\partial}{\partial n_l} \left\{ t_s^2 + \lambda \left(n_k n_k - 1 \right) \right\} = 0$$

yields

$$n_l \left\{ \sigma_l^2 - 2\sigma_l t_n + \lambda \right\} = 0 \text{ (no sum on } l)$$

where λ is the Lagrange multiplier. Writing out the three equations for $l = 1, 2, 3$ gives

$$n_1 \left\{ \sigma_1^2 - 2\sigma_1 t_n + \lambda \right\} = 0 \qquad (11.4)$$

$$n_2 \left\{ \sigma_2^2 - 2\sigma_2 t_n + \lambda \right\} = 0 \qquad (11.5)$$

$$n_3 \left\{ \sigma_3^2 - 2\sigma_3 t_n + \lambda \right\} = 0 \qquad (11.6)$$

There are three possible cases corresponding to one, two, or none of the n_l being zero.

Case 1: Suppose, for example, that $n_2 = n_3 = 0$; then $n_1 = 1$. Equations (11.5) and (11.6) are automatically satisfied. The only non-trivial equation (11.4) reduces to

$$\sigma_1^2 - 2\sigma_1 t_n + \lambda = 0$$

For $n_2 = n_3 = 0$, $n_1 = 1$, $t_n = \sigma_1$, and, consequently, $\lambda = \sigma_1^2$. Substituting back into (11.1) yields $t_s^2 = 0$. Because $\mathbf{n}_1 = \mathbf{e}_1$ is a principal direction, this result simply confirms that the shear traction is zero on principal planes.

Case 2: Now suppose that $n_1, n_2 \neq 0$ and $n_3 = 0$. Equation (11.6) is automatically satisfied and (11.4) and (11.5) become

$$\sigma_1^2 - 2\sigma_1 t_n + \lambda = 0$$

$$\sigma_2^2 - 2\sigma_2 t_n + \lambda = 0$$

Eliminating λ and rearranging gives

$$(\sigma_1 - \sigma_2)(\sigma_1 + \sigma_2) = 2(\sigma_1 - \sigma_2)t_n$$

Assuming $\sigma_1 \neq \sigma_2$, writing

$$t_n = (n_1^2 \sigma_1 + n_2^2 \sigma_2)$$

and substituting $n_2^2 = 1 - n_1^2$ yield

$$n_1^2 = \frac{1}{2} \quad \text{or} \quad n_1 = \pm \frac{1}{\sqrt{2}}$$

and, consequently,

$$n_2 = \pm \frac{1}{\sqrt{2}}$$

Taking the plus signs and substituting into expression (11.1) gives

$$t_s^2 = \sum_k n_k^2 \sigma_k^2 - \left[\sum_k (n_k \sigma_k n_k) \right]^2$$

$$= \frac{1}{2}(\sigma_1^2 + \sigma_2^2) - \frac{1}{4}(\sigma_1 + \sigma_2)^2 = \frac{1}{4}(\sigma_1 - \sigma_2)^2$$

or

$$(t_s)_{\max} = \frac{1}{2}|\sigma_1 - \sigma_2| \qquad (11.7)$$

Case 3: If all $n_l \neq 0$, then the principal stresses cannot be distinct. If two of the principal stresses are equal, the normal to the plane of greatest shear traction can be within a cone at 45° to the principal direction of the distinct principal stress. Because any two directions within the plane of the equal principal stresses can be principal directions, it is always possible to choose them so that one of the n_i can be zero. Thus this case effectively reduces to Case 2. If all of the principal stresses are equal, $t_s^2 \equiv 0$ on all planes.

Equation (11.7) gives the maximum shear traction on planes with normals in the 1 and 2 planes. The same calculation yields corresponding results for normals in the 1 and 3 and 2 and 3 planes. Therefore the absolute maximum value of t_s occurs on a plane with a normal that makes a 45° angle with the principal directions corresponding to the maximum and minimum principal stresses.

The derivation here has been for the stress tensor, but the same results apply for symmetric tensors with other physical interpretations.

11.1   Example: Mohr–Coulomb Failure Condition

The Mohr–Coulomb criterion is a common failure condition for geomaterials loaded in compression. The criterion states that failure occurs when the magnitude of the shear traction on a plane t_s is equal to the cohesion τ_0 minus a friction coefficient μ multiplied by the normal traction on that plane t_n ($t_n < 0$ for compression):

$$|t_s| = \tau_0 - \mu t_n \qquad (11.8)$$

What is the orientation of the plane on which this criterion is first met?

Let the axes coincide with the principal stress directions so that the stress tensor is given by

$$\sigma = -\sigma_1 e_1 e_1 - \sigma_2 e_2 e_2 - \sigma_3 e_3 e_3$$

where $\sigma_3 > \sigma_2 > \sigma_1 > 0$. Because the maximum shear stress occurs in the $e_1 e_3$ plane, we assume that the normal to the plane of interest is also in this plane. Let the normal to this plane make an angle α with e_1:

$$n = \cos(\alpha) e_1 + \sin(\alpha) e_2$$

and the tangent be

$$\mathbf{s} = -\sin{(\alpha)}\,\mathbf{e}_1 + \cos{(\alpha)}\,\mathbf{e}_2$$

The normal component of the traction is

$$t_n = -\frac{1}{2}(\sigma_1 + \sigma_3) + \frac{1}{2}(\sigma_3 - \sigma_1)\cos(2\alpha) \qquad (11.9)$$

and the shear component is

$$t_s = -\frac{1}{2}(\sigma_3 - \sigma_1)\sin(2\alpha) \qquad (11.10)$$

where we have used the double angle formulae

$$\cos^2{(\alpha)} = \frac{1}{2}(1 + \cos 2\alpha)$$

$$\sin^2{(\alpha)} = \frac{1}{2}(1 - \cos 2\alpha)$$

Note that

$$|t_s| = \begin{cases} \dfrac{1}{2}(\sigma_3 - \sigma_1)\sin(2\alpha), & 0 \le \alpha \le \pi/2 \\[2mm] -\dfrac{1}{2}(\sigma_3 - \sigma_1)\sin(2\alpha), & -\pi/2 \le \alpha \le 0 \end{cases}$$

Setting $\mu = \tan\phi$, substituting (11.10), and (11.9) into (11.8), and rearranging gives

$$\frac{1}{2}(\sigma_3 - \sigma_1)\frac{\sin{(\phi \pm 2\alpha)}}{\cos\phi} = \tau_0 + \frac{1}{2}\tan\phi\,(\sigma_1 + \sigma_3)$$

Thus, the smallest value of $(\sigma_3 - \sigma_1)/2$ that meets the criterion occurs at the orientation α that makes $\sin{(\phi \pm 2\alpha)}$ largest; that is, when

$$\frac{\partial}{\partial\alpha}\sin{(\phi \pm 2\alpha)} = 0$$

or

$$\alpha = \pm\left(\frac{\pi}{4} - \frac{\phi}{2}\right)$$

Substituting back into (11.8) gives the relation between the principal stresses σ_1 and σ_3 at failure:

$$\frac{1}{2}\sigma_3\,(1 - \sin\phi) = \frac{1}{2}\sigma_1\,(1 + \sin\phi) + \tau_0\cos\phi$$

Exercises

11.1 Carry out the details of Case 3 to show that if all $n_l \neq 0$, then the principal stresses cannot be distinct.

11.2 The components of the stress tensor at a point are given by

$$[\sigma] = \begin{bmatrix} -1 & -2 & 0 \\ -2 & 0 & 2 \\ 0 & 2 & 1 \end{bmatrix}$$

(a) Determine the traction on a plane with unit normal $\mathbf{n} = (\mathbf{e}_1 + 2\mathbf{e}_2 + 2\mathbf{e}_3)/3$.
(b) Determine the magnitude of the normal traction and the magnitude of the shear traction on this plane.

11.3 The stress tensor is given by

$$\sigma = T\mathbf{uu}$$

where $\mathbf{u}$ is a unit vector and $T > 0$.
(a) Determine the traction on a plane with normal that makes an angle θ with $\mathbf{u}$.
(b) Determine the normal value of the traction on this plane.
(c) Determine the magnitude of the tangential component of the traction on this plane.

11.4 For the stress tensor given in Problem 10.2 determine the maximum shear stress and the *normal to the plane* on which it occurs.

11.5 Show that the shear traction vanishes on every plane if and only if the stress tensor is given by $\sigma = -p\mathbf{I}$, where p is a scalar and $\mathbf{I}$ is the identity tensor.

11.6 Let the stress tensor be given by $\sigma = \sigma_1\mathbf{e}_1\mathbf{e}_1 + \sigma_2\mathbf{e}_2\mathbf{e}_2 + \sigma_3\mathbf{e}_3\mathbf{e}_3$.
(a) Determine the traction vector on a plane with a normal that makes equal angles with the coordinate directions.
(b) Determine the normal component of the traction on this plane.
(c) Show that the magnitude of the shear component can be expressed as

$$t_s = \frac{1}{3}\left\{(\sigma_1 - \sigma_2)^2 + (\sigma_2 - \sigma_3)^2 + (\sigma_3 - \sigma_1)^2\right\}^{1/2}$$

and, hence, $J_2 = (3/2)\,t_s^2$ from Problem 10.3.
(d) Determine the direction of the shear traction.

11.7 Let the principal stresses at a point satisfy the relation $\sigma_2 = \frac{1}{2}(\sigma_1 + \sigma_3) > 0$. Determine the orientation of the plane (relative to the principal axes) on which the normal traction is $t_n = \sigma_2$ and the shear (tangential) traction is $t_s = (\sigma_1 - \sigma_3)/4$.

12

Mohr's Circle

Mohr's circle is a graphical construction that is familiar to students of mechanics of materials. Although it has long outlived its usefulness as a computational aid, it provides a physical illustration of the meaning of a tensor, in particular the stress tensor. In addition, it is a point of contact between the present treatment and mechanics of materials. In three dimensions, it is most useful when at least one of the principal values of the stress is already known. Here, the x_3 direction is assumed to be a principal direction (Figure 12.1) for the stress tensor. Consequently, the stress tensor can be written as

$$\sigma = \sigma_{\gamma\delta}\mathbf{e}_\gamma\mathbf{e}_\delta + \sigma_{III}\mathbf{e}_3\mathbf{e}_3$$

where $\gamma, \delta = 1, 2$.

Now consider the traction components on a plane with normal $\mathbf{n}$ making an angle α with the x_1 axis as shown in Figure 12.2. The normal and tangent vectors are

$$\mathbf{n} = \cos\alpha\mathbf{e}_1 + \sin\alpha\mathbf{e}_2 \tag{12.1}$$

$$\mathbf{s} = -\sin\alpha\mathbf{e}_1 + \cos\alpha\mathbf{e}_2 \tag{12.2}$$

When $\alpha = 0°$, $t_n = \sigma_{11}$ and $t_s = \sigma_{12}$, and when $\alpha = 90°$, $t_n = \sigma_{22}$ and $t_s = -\sigma_{12}$. This means that shear stress components tending to cause a clockwise moment are plotted as negative in Mohr's circle (even though the shear stress components themselves may be positive). This difference in sign results from the difference between the component of the traction, which is a vector, and the component of the stress, which is a tensor. Alternatively, we could have taken the positive $\mathbf{s}$ direction to be clockwise from $\mathbf{n}$, in which case the signs on the shear traction would be reversed. This choice governs whether the rotation in the Mohr plane, which plots t_s against t_n, is in the same or the opposite direction as the rotation in the physical plane. (Malvern (1988) describes both conventions.)

Fundamentals of Continuum Mechanics, First Edition. John W. Rudnicki.
© 2015 John Wiley & Sons, Ltd. Published 2015 by John Wiley & Sons, Ltd.

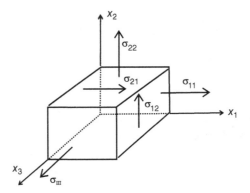

Figure 12.1 Element for analysis with Mohr's circle. The x_3 direction is a principal direction.

The traction vector on the inclined plane in Figure 12.2 is

$$\mathbf{t} = \mathbf{n} \cdot \sigma = \cos \alpha (\sigma_{11} \mathbf{e}_1 + \sigma_{12} \mathbf{e}_2) + \sin \alpha (\sigma_{21} \mathbf{e}_1 + \sigma_{22} \mathbf{e}_2)$$

The normal component is

$$t_n = \mathbf{n} \cdot \mathbf{t} = \frac{1}{2}(\sigma_{11} + \sigma_{22}) - \frac{1}{2}(\sigma_{22} - \sigma_{11}) \cos 2\alpha + \sigma_{12} \sin 2\alpha$$

and the shear component is

$$t_s = \mathbf{s} \cdot \mathbf{t} = \frac{1}{2}(\sigma_{22} - \sigma_{11}) \sin 2\alpha + \sigma_{12} \cos 2\alpha$$

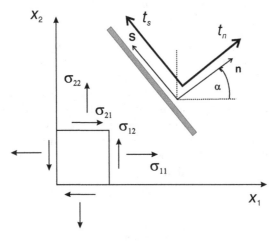

Figure 12.2 Normal, t_n, and shear, t_s, tractions on an inclined plane with normal **n**.

where we have used the double angle formulae:

$$\cos^2 \theta = \frac{1}{2}(1 + \cos 2\theta)$$

$$\sin^2 \theta = \frac{1}{2}(1 - \cos 2\theta)$$

$$\sin 2\theta = 2 \cos \theta \sin \theta$$

Forming

$$\left[\left\{ t_n - \frac{1}{2}(\sigma_{11} + \sigma_{22}) \right\}^2 + t_s^2 \right] = R^2$$

gives the equation of a circle in the plane t_s vs. t_n (Figure 12.3). The center of the circle is at $t_n = \frac{1}{2}(\sigma_{11} + \sigma_{22})$ and the radius is

$$R = \sqrt{\left\{ \frac{1}{2}(\sigma_{11} - \sigma_{22}) \right\}^2 + (\sigma_{12})^2}$$

The points on the circle give the values of t_s and t_n as the angle α varies. Because these all originate from the same stress state shown in Figure 12.1, the circle is a graphical representation of a tensor in two dimensions. The center and radius are invariants of the two-dimensional stress state. The traction on the plane perpendicular to the x_1 direction is represented by the point $(\sigma_{11}, \sigma_{12})$ corresponding to $\alpha = 0°$, and the traction on the plane perpendicular to the x_2 direction is represented by the point $(\sigma_{22}, -\sigma_{21})$ corresponding to $\alpha = 90°$. (The points are depicted in Figure 12.3 assuming $\sigma_{11} > \sigma_{22}$ and $\sigma_{12} > 0$.) Although $\sigma_{12} = \sigma_{21}$, the minus sign is necessary for the point corresponding to $\theta = 90°$ because, when the plane in Figure 12.2 is oriented perpendicular to the x_2 direction, s and σ_{21} point in opposite directions. As already noted, this feature occurs because Mohr's circle is a plot of the components of traction, a vector, whereas stress is a tensor.

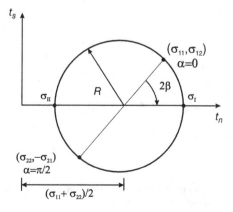

Figure 12.3 Mohr's circle.

It is obvious from the geometry of the circle that the largest and smallest values of the normal traction are the principal stresses $\sigma_{I,II} = \frac{1}{2}(\sigma_{11} + \sigma_{22}) \pm R$ and that these occur on planes where the shear traction is zero, $t_s = 0$. The points corresponding to the principal stresses lie at the opposite ends of a diameter, 180° apart in the Mohr plane. We know from Chapter 7 and Chapter 10 that the principal stresses occur on orthogonal planes. Also, the magnitude of the shear traction is greatest at the points at the top and bottom of the circle. These points are 90° from the points corresponding to the principal stresses in the Mohr plane. In Chapter 10 we showed that the maximum magnitude of shear traction occurs on planes 45° from the principal planes and these are represented by points 90° from the points in the Mohr plane representing the principal planes.

These observations make it clear that planes with normals α apart in the physical plane are represented by points 2α apart in the Mohr plane. Consequently, the angle between the x_1 direction and the normal to the principal plane on which σ_I acts is

$$\beta = \frac{1}{2} \arctan \left(\frac{2\sigma_{12}}{\sigma_{11} - \sigma_{22}} \right)$$

The question that remains is whether the rotation in the Mohr plane is in the same sense as in the physical plane. This depends on whether the positive direction for vector **s** is taken to be a clockwise or counterclockwise rotation from the direction of the normal **n**. For the counterclockwise rotation for **s**, as assumed in Figure 12.1, rotations in the Mohr plane correspond to rotations in the opposite sense in the physical plane. In other words, a clockwise rotation in the Mohr plane corresponds to a counterclockwise rotation of the normal in the physical plane. This can be confirmed by using the methods of Chapter 7 to find orientations of the principal directions.

Mohr's circle also can be used to visualize the change in stress components due to a rotation of the coordinate axes. Taking $\alpha = \theta$ and $\alpha = \theta + 90°$ in Figure 12.2 locates points corresponding to the stress components for a rotation of coordinate axes through an angle θ. This is shown in Figure 12.4.

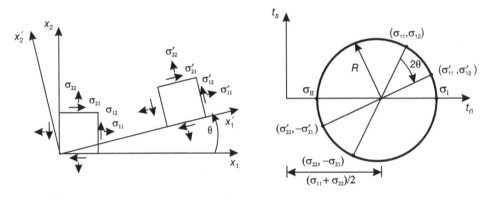

Figure 12.4   Illustration of Mohr's circle representation of the change in stress components due to a rotation of axes through a counterclockwise angle θ.

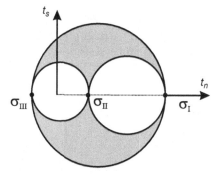

Figure 12.5 Schematic of the Mohr's circles for the planes of each pair of principal stresses. All possible values of the traction components t_s and t_n lie in the shaded area between the circles.

Figure 12.3 illustrates how the traction and stress components vary on orientations in the plane of σ_I and σ_{II}. Similar constructions apply for the planes of σ_{II} and σ_{III} and σ_I and σ_{III}. Thus, there are three circles as depicted in Figure 12.5. The possible values of the traction components t_s and t_n occupy the shaded region between the circles.

Exercises

12.1 For each of the following stress states (values not given are zero), sketch the three Mohr's circles. What is the maximum shear stress in each case, and what is the normal stress on the plane of maximum shear stress (a and σ are positive constants)?
   (a) Uniaxial compression, $\sigma_{11} = -\sigma$.
   (b) Biaxial stress, $\sigma_{11} = +a$, $\sigma_{22} = -3a$.
   (c) Hydrostatic compression of magnitude σ.
   (d) $\sigma_{12} = \sigma_{21} = 3a$, $\sigma_{23} = \sigma_{32} = 4a$.
   (e) $\sigma_{11} = -5a$, $\sigma_{22} = -a$, $\sigma_{33} = a$.

12.2 Discuss the position of the point corresponding to the plane in Problem 11.7 in relation to the Mohr's circles.

12.3 Use Mohr's circle to obtain the result of Example 11.1.

Reference

Malvern LE 1988 *Introduction to the Mechanics of a Continuous Medium*. Prentice Hall.

Part Three

Motion and Deformation

In this part we develop mathematical descriptions of the geometry of motion and deformation. The descriptions are purely kinematic; that is, we do not consider the forces that give rise to the motion and deformation. The descriptions will not make assumptions about the magnitude of the deformations, but there are various possibilities of description that are more convenient for different types of problems. To this end, we will introduce a number of tensors and make use of the material in Chapter 2, Chapter 3, and, for symmetric tensors, Chapter 7.

Fundamentals of Continuum Mechanics, First Edition. John W. Rudnicki.
© 2015 John Wiley & Sons, Ltd. Published 2015 by John Wiley & Sons, Ltd.

13

Current and Reference Configurations

Figure 13.1 shows two configurations of an arbitrary body: The *reference configuration* at some time t_0 and the *current configuration* at time $t \geq t_0$. The reference configuration can be chosen for convenience in analysis. For example, for an elastic body, it is usually convenient to choose the reference configuration as the configuration when the loads are reduced to zero. For an elastic–plastic body or a fluid, it is often convenient to choose the reference configuration to coincide with the current configuration and to focus on increments or rates from the current configuration.

In Figure 13.1 $P_0(\mathbf{X})$ is the position of a material particle in the reference configuration. The same material particle is located at $P(\mathbf{x})$ in the current configuration. Here, positions in both the reference configuration and the current configuration are referred to the same rectangular Cartesian coordinate system. This is not necessary and, often, it is more convenient to refer positions in the two configurations to different coordinate systems. The *motion* of the material particle is described by

$$\mathbf{x} = \phi(\mathbf{X}, t) \tag{13.1}$$

or

$$\mathbf{x} = \phi(X_1, X_2, X_3, t) \tag{13.2}$$

and is usually abbreviated

$$\mathbf{x} = \mathbf{x}(\mathbf{X}, t) \tag{13.3}$$

The notation in (13.3), although ubiquitous, can be confusing. In (13.1) or (13.2) $\mathbf{x}$ is used to denote the value of the function ϕ for a particular $\mathbf{X}$ and t. In (13.3) $\mathbf{x}$ denotes both the function ϕ and its value for a particular $\mathbf{X}$ and t. In words, these expressions say that "$\mathbf{x}$ is the position at time t of the particle that occupied position $\mathbf{X}$ in the reference configuration at time $t = t_0$." In this description $\mathbf{x}$ is regarded as the dependent variable; $\mathbf{X}$ is the independent variable. Because each material particle occupies a unique position in the reference configuration, the position $\mathbf{X}$ can be used as a label for the particle. That is, different values of $\mathbf{X}$ correspond to different

Fundamentals of Continuum Mechanics, First Edition. John W. Rudnicki.
© 2015 John Wiley & Sons, Ltd. Published 2015 by John Wiley & Sons, Ltd.

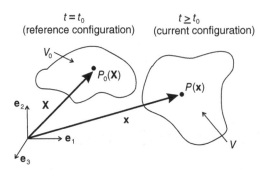

Figure 13.1   Schematic of the reference and current configurations.

material particles. The position **x** may, however, be occupied by different material particles at different times.

An analogy is to consider students as material particles. Rather than referring to them by name, student ID, social security number, etc., we can refer to them by their position at a particular time, say the class period 9 to 10 a.m. Thus, each student is labeled by the position of his or her seat in the class, for example, the third seat from the left in the second row. This is the reference configuration and the seat is **X**. After the class period the students go about their business moving to other locations, but we continue to refer to them by the location of their seat during the class period. At some time t after the class period, say 2 p.m., the student who occupied the third seat from the left in the second row in the reference configuration will be at a position **x** in the student union. During other times of the day, different students will occupy that position. Hence an observer standing in the student union would see different students (material particles) occupy the position **x** at different times.

Physically, it is plausible that the motion can be inverted because each and every point in the reference configuration corresponds to exactly one point in the current configuration. Therefore, at least in principle, we can invert the motion to write the position in the reference configuration **X** in terms of time and the current location **x**:

$$\mathbf{X} = \Phi(\mathbf{x}, t) \quad \text{or} \quad \mathbf{X} = \mathbf{X}(\mathbf{x}, t) \tag{13.4}$$

Now we regard **x** as the independent variable. The mathematical condition insuring that (13.1) can be inverted is

$$J = \left| \frac{\partial \mathbf{x}}{\partial \mathbf{X}} \right| = \left| \frac{\partial x_i}{\partial X_j} \right| > 0 \tag{13.5}$$

The left side is the determinant of a particular tensor to be introduced in the next chapter. Because the determinant of this tensor does not vanish, its inverse exists as discussed following (5.8). We will show that (13.5) expresses the physical requirement that small volume elements in both the reference and current configurations are finite and positive.

When **X** is used as the independent variable, this is often called the *Lagrangian* description. Because different values of **X** correspond to different positions in the reference configuration and hence different material particles, the Lagrangian description follows a material particle through the motion. A physical example is following the motion of a radioactive particle or a group of particles marked with dye.

The *Eulerian* description uses $\mathbf{x}$ as independent variable. This point of view considers a fixed location in space and observes how the material particles move past this location. Because a fixed value of $\mathbf{x}$ refers to a fixed location, it does not correspond to a particular material particle; that is, different particles will move past this location as time evolves. A physical example is measurement by an instrument at a fixed location as different particles move past.

If the motion, (13.1), (13.2), or (13.3), is known, the velocity can be computed simply as the rate of change of the location with time:

$$\mathbf{V}(\mathbf{X}, t) = \frac{\partial \mathbf{x}}{\partial t} = \frac{\partial \boldsymbol{\phi}}{\partial t}(\mathbf{X}, t) \tag{13.6}$$

Because $\partial/\partial t$ means to take the derivative with respect to time while holding the other arguments, i.e., $\mathbf{X}$, fixed, (13.6) gives an expression for the velocity of the particle that was located at $\mathbf{X}$ at time t_0. (Note that this particle is not now, at time t, located at $\mathbf{X}$.) Thus, (13.6) is the Lagrangian description of the velocity. We will use upper case letters to denote quantities given in terms of the Lagrangian description.

To get the Eulerian description, we substitute (13.4) into the argument of (13.6)

$$\mathbf{v}(\mathbf{x}, t) = \mathbf{V}\left[\boldsymbol{\Phi}(\mathbf{x}, t), t\right] \tag{13.7}$$

Note that if $\mathbf{x}$ is the current position of the particular particle that was located at $\mathbf{X}$ in the reference configuration then the values of the velocity given by (13.6) and (13.7) must be equal.

Now, consider any scalar property θ, e.g., temperature, density. The Eulerian description is

$$\theta = \theta(\mathbf{x}, t) \tag{13.8}$$

and the Lagrangian description is

$$\Theta = \Theta(\mathbf{X}, t) \tag{13.9}$$

The partial derivative of (13.8)

$$\left. \frac{\partial \theta}{\partial t} \right|_{\mathbf{x}\,\text{fixed}}$$

gives the rate of change of θ at a fixed location in space. This is *not* the rate of change of θ of any material particle because different particles occupy the location $\mathbf{x}$ as time t changes. The partial derivative of (13.9)

$$\left. \frac{\partial \Theta}{\partial t} \right|_{\mathbf{X}\,\text{fixed}} \tag{13.10}$$

does give the rate of change of Θ for a specific material particle.

Can we compute the rate of change of Θ for a material particle if we are given only $\theta(\mathbf{x}, t)$? Mathematically, this can be expressed as follows:

$$\frac{\partial \Theta}{\partial t}(\mathbf{X}, t) = \left. \frac{d\theta}{dt} \right|_{\mathbf{X}\,\text{fixed}}$$

Because the right hand side is evaluated for fixed $\mathbf{X}$, the location of the particle, $\mathbf{x}$, changes with time. Therefore, by the chain rule of differentiation,

$$\left. \frac{d\theta}{dt} \right|_{\mathbf{X}\,\text{fixed}} = \left. \frac{\partial \theta}{\partial t}(\mathbf{x}, t) \right|_{\mathbf{x}\,\text{fixed}} + \frac{\partial \theta}{\partial x_i} \frac{\partial x_i}{\partial t} \qquad (13.11)$$

Note that $\partial x_i/\partial t$ is the component form of the velocity of a particle and $\partial \theta/\partial x_i$ are the components of the gradient of θ. Thus (13.11) can be written in coordinate-free vector form as

$$\frac{d\theta}{dt} = \left. \frac{d\theta}{dt} \right|_{\mathbf{X}\,\text{fixed}} = \left. \frac{\partial \theta}{\partial t} \right|_{\mathbf{x}\,\text{fixed}} + \mathbf{v}(\mathbf{x}, t) \cdot \nabla\theta \qquad (13.12)$$

Equation (13.12) gives the rate of change of θ following a material particle or the *material derivative*. The designation $\mathbf{X}$ fixed is usually omitted and to be understood. The derivative (13.12) is the same as what is called the *total derivative* in calculus but here has the specific meaning of following a material particle. Because holding $\mathbf{x}$ fixed in the first term on the right corresponds to the usual meaning of the partial derivative, the notation explicitly indicating that $\mathbf{x}$ is fixed is usually omitted. Note that in order to compute the material rate of change of $\theta(\mathbf{x}, t)$, it is necessary to know not only the change of θ at a particular location $\mathbf{x}$, $\partial \theta/\partial t$, but also how θ is changing in an infinitesimal interval about $\mathbf{x}$, $\partial \theta/\partial x_i$. The expressions (13.10) and (13.12) must give the same value if they are evaluated for the same particle at the same time, regardless of whether the particle is specified by its current location or its location in the reference configuration.

Similarly, the material rate of change can also be computed for a vector property $\mu(\mathbf{x}, t)$:

$$\frac{d\mu}{dt}(\mathbf{x}, t) = \left. \left(\frac{\partial \mu}{\partial t} \right) \right|_{\mathbf{x}\,\text{fixed}} + \mathbf{v} \cdot (\nabla \mu)$$

If $\mu = v$, the velocity, then the material derivative gives the Eulerian description of the acceleration

$$\mathbf{a}(\mathbf{x}, t) = \frac{d\mathbf{v}}{dt}(\mathbf{x}, t) = \frac{\partial \mathbf{v}}{\partial t} + \mathbf{v} \cdot \nabla \mathbf{v}$$

The Lagrangian description of the acceleration is

$$\mathbf{A}(\mathbf{X}, t) = \frac{\partial \mathbf{V}(\mathbf{X}, t)}{\partial t}$$

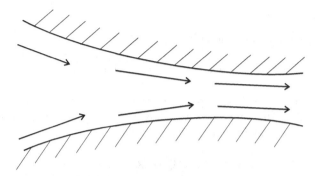

Figure 13.2 Example of steady flow of an incompressible fluid down a converging channel. Because the flow is steady, the velocity does not change at any fixed location, $\partial \mathbf{v}/\partial t = 0$. But because particles increase their velocity as they move down the channel, the acceleration is nonzero.

Flow of an incompressible fluid down a converging channel illustrates the difference between the material derivative d/dt and $\partial/\partial t$. In Figure 13.2 the flow is steady, meaning that $\partial \mathbf{v}/\partial t = 0$ because the velocity does not change at any fixed location. But the acceleration $d\mathbf{v}/dt \neq 0$ because material particles increase their velocity as they move down the channel. In Figure 13.3 the fluid is initially at rest. Then the fan is turned on. Consequently, the velocity of (different) particles passing a fixed location changes with time, $\partial \mathbf{v}/\partial t \neq 0$.

When does $\partial \theta/\partial t = d\theta/dt$ for a property θ? This will be true if the second term in (13.12) vanishes

$$\mathbf{v} \cdot \nabla \theta = 0$$

There are three possibilities: (i) $\mathbf{v} = 0$ so that there is no motion; (ii) $\nabla \theta = 0$ so that θ is spatially uniform; (iii) $\mathbf{v}$ is perpendicular to $\nabla \theta$. An example of (iii) is a motion in which the only nonzero component of velocity is in the x_1 direction, $\mathbf{v} = v_1 \mathbf{e}_1$, θ is temperature, and there is a gradient of temperature only in the x_2 direction.

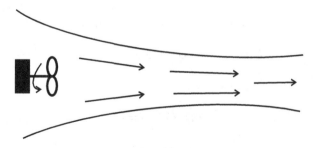

Figure 13.3 In this example, the material is initially at rest and then the fan is turned on. Consequently, the velocity is changing at fixed spatial locations and $\partial \mathbf{v}/\partial t \neq 0$.

13.1   Example

The motion of a rigid body can be described by

$$\mathbf{x}(\mathbf{X}, t) = \mathbf{Q}(t) \cdot \mathbf{X} + \mathbf{c}(t) \tag{13.13}$$

Determine expressions for the Lagrangian descriptions of velocity and acceleration. Also determine expressions for the Eulerian descriptions of velocity and acceleration.

The Lagrangian velocity and acceleration are simply

$$\mathbf{V}(\mathbf{X}, t) = \dot{\mathbf{Q}}(t) \cdot \mathbf{X} + \dot{\mathbf{c}}(t) \tag{13.14}$$

and

$$\mathbf{A}(\mathbf{X}, t) = \ddot{\mathbf{Q}}(t) \cdot \mathbf{X} + \ddot{\mathbf{c}}(t) \tag{13.15}$$

To obtain the Eulerian descriptions of velocity and acceleration, we solve (13.13) for

$$\mathbf{X}(\mathbf{x}, t) = \mathbf{Q}^{-1}(t) \cdot \{\mathbf{x} - \mathbf{c}(t)\}$$

and substitute into (13.14) and (13.15) to get

$$\mathbf{v}(\mathbf{x}, t) = \dot{\mathbf{Q}}(t) \cdot \mathbf{Q}^{-1}(t) \cdot \{\mathbf{x} - \mathbf{c}(t)\} + \dot{\mathbf{c}}(t)$$

and

$$\mathbf{a}(\mathbf{x}, t) = \ddot{\mathbf{Q}}(t) \cdot \mathbf{Q}^{-1}(t) \cdot \{\mathbf{x} - \mathbf{c}(t)\} + \ddot{\mathbf{c}}(t) \tag{13.16}$$

The Eulerian description of acceleration can also be obtained by using the material derivative:

$$\mathbf{a}(\mathbf{x}, t) = \frac{\partial}{\partial t}\mathbf{v} + \mathbf{v} \cdot \nabla\mathbf{v} \tag{13.17}$$

The first term is

$$\frac{\partial}{\partial t}\mathbf{v} = \left\{ \ddot{\mathbf{Q}}(t) \cdot \mathbf{Q}^{-1}(t) + \dot{\mathbf{Q}}(t) \cdot \frac{\partial}{\partial t}\left(\mathbf{Q}^{-1}(t)\right) \right\} \cdot \{\mathbf{x} - \mathbf{c}(t)\}$$
$$- \dot{\mathbf{Q}}(t) \cdot \mathbf{Q}^{-1}(t) \cdot \dot{\mathbf{c}}(t) + \ddot{\mathbf{c}}(t) \tag{13.18}$$

To calculate $\partial(\mathbf{Q}^{-1}(t))/\partial t$, we differentiate

$$\mathbf{Q}(t) \cdot \mathbf{Q}^{-1}(t) = \mathbf{I}$$

and form the tensor product of each side with $\mathbf{Q}^{-1}(t)$ to obtain

$$\frac{\partial}{\partial t}(\mathbf{Q}^{-1}(t)) = -\mathbf{Q}^{-1}(t) \cdot \dot{\mathbf{Q}}(t) \cdot \mathbf{Q}^{-1}(t) \tag{13.19}$$

Noting that

$$\nabla \mathbf{v} = \dot{\mathbf{Q}}(t) \cdot \mathbf{Q}^{-1}(t) \tag{13.20}$$

and substituting (13.20) into (13.17) along with the result of substituting (13.19) into (13.18) yields (13.16).

Exercises

13.1 **(a)** Determine the motion that deforms the unit square to the rhombus shown in Figure 13.4, where $\gamma = \gamma(t)$.
  (b) Determine the Lagrangian and Eulerian descriptions of the velocity.
  (c) Invert the motion to determine the position in the reference configuration in terms of the current position.

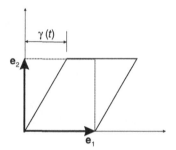

Figure 13.4 Simple shear deformation

13.2 The components of the velocity in a material are given by $v_i = x_i/(1 + kt)$ and the temperature is given by $\theta = x_1 + ktx_2$ where t is time and k is a constant with dimension $(\text{time})^{-1}$.
  (a) Calculate the components of the acceleration.
  (b) Calculate the material rate of change of the temperature.

13.3 Consider the motion

$$\mathbf{x} = \mathbf{X}(1 + kt)$$

where t is time and k is a constant with dimension $(\text{time})^{-1}$.
  (a) Determine the Eulerian (spatial) and Lagrangian (material) descriptions of the velocity.
  (b) Determine the Lagrangian (material) description of the acceleration.
  (c) Determine the Eulerian (spatial) description of the acceleration in two ways: By taking the material derivative of your answer from (b); and by inverting the motion and using your answer in (a).

13.4 Consider the motion

$$x_1 = X_1 + X_2(e^t - 1)$$
$$x_2 = X_1(e^{-t} - 1) + X_2$$
$$x_3 = X_3$$

(a) Determine the expressions for the Lagrangian description of velocity and acceleration.

(b) Determine the expressions for the Eulerian description of velocity.

(c) Determine the Eulerian description of the acceleration in two ways: By inverting the motion and using your answer from (a); and by taking the material derivative of your answer in (b).

14

Rate of Deformation

14.1 Velocity Gradients

In some cases, the reference configuration is not of interest. We are not concerned with the locations of particles at some past time, but only with the instantaneous velocity field. For example, in the flow of a fluid, a configuration at a past time is generally not useful (or even possible to identify). In other cases, the past location of particles is of interest but the response depends on the history of deformation. Consequently, the solution needs to be determined incrementally, i.e., step by step. This corresponds to taking the reference configuration as instantaneously coincident with the current configuration and updating it at each increment.

Consider a velocity field $\mathbf{v}(\mathbf{x})$ as shown in Figure 14.1. Although the particles were at points P and Q in the reference configuration, we are interested only in the instantaneous velocities of these points at their current locations p and q. The difference between the velocity of a particle located at $\mathbf{x}$ and a particle located at $\mathbf{x} + d\mathbf{x}$ at the current time is

$$d\mathbf{v} = \mathbf{v}(\mathbf{x} + d\mathbf{x}) - \mathbf{v}(\mathbf{x})$$

or, in component form,

$$dv_k = v_k(\mathbf{x} + d\mathbf{x}) - v_k(\mathbf{x}) = \frac{\partial v_k}{\partial x_l} dx_l \tag{14.1}$$

where t has been omitted as an argument. The last equality in (14.1) can be rationalized by expanding $v_k(\mathbf{x} + d\mathbf{x})$ in a Taylor series and retaining only first terms (as we did earlier in determining the form for the gradient of a vector, (8.4) to (8.7)). We can write this result in coordinate-free form as

$$d\mathbf{v} = \mathbf{L} \cdot d\mathbf{x} = d\mathbf{x} \cdot \mathbf{L}^T \tag{14.2}$$

where

$$\mathbf{L} = (\nabla v)^T \tag{14.3}$$

Fundamentals of Continuum Mechanics, First Edition. John W. Rudnicki.
© 2015 John Wiley & Sons, Ltd. Published 2015 by John Wiley & Sons, Ltd.

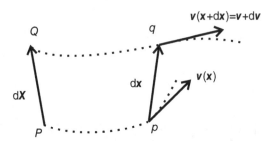

Figure 14.1 Illustration of the velocity difference at points p and q in the current configuration that are separated by an infinitesimal distance d**x**.

and is given in component form by

$$L_{kl} = \frac{\partial v_k}{\partial x_l} = v_{k,l}$$

L is the (spatial) velocity gradient tensor. The symmetric part of **L**

$$\mathbf{D} = \frac{1}{2}(\mathbf{L} + \mathbf{L}^T) \tag{14.4}$$

is the rate-of-deformation tensor and the anti- or skew-symmetric part of **L**

$$\mathbf{W} = \frac{1}{2}(\mathbf{L} - \mathbf{L}^T) \tag{14.5}$$

is the spin tensor or vorticity tensor.

14.2   Meaning of D

The meaning of **D** can be established by considering the rate of change of the length of an infinitesimal line segment d**x**. The length squared is

$$ds^2 = d\mathbf{x} \cdot d\mathbf{x}$$

Differentiating with respect to time gives

$$2\, ds \frac{d}{dt}(ds) = d\mathbf{v} \cdot d\mathbf{x} + d\mathbf{x} \cdot d\mathbf{v}$$

where we have used

$$\frac{d}{dt}(d\mathbf{x}) = d\left(\frac{d\mathbf{x}}{dt}\right) = d\mathbf{v}$$

Using (14.2) gives

$$2\,ds\frac{d}{dt}(ds) = d\mathbf{x} \cdot \mathbf{L}^T \cdot d\mathbf{x} + d\mathbf{x} \cdot \mathbf{L} \cdot d\mathbf{x} = 2\,d\mathbf{x} \cdot \mathbf{D} \cdot d\mathbf{x}$$

Dividing by $2\,ds^2$ yields

$$\frac{1}{ds}\frac{d}{dt}(ds) = \frac{d}{dt}\ln(ds) = \mathbf{n} \cdot \mathbf{D} \cdot \mathbf{n}$$

Thus, $\mathbf{n} \cdot \mathbf{D} \cdot \mathbf{n}$ is the fractional rate of extension in direction $\mathbf{n} = d\mathbf{x}/ds$. Normal components of $\mathbf{D}$ give the fractional rates of extension of line segments in the coordinate directions. Since $\mathbf{D}$ is a symmetric tensor, it has three real principal values with orthogonal principal directions. The same derivation used in Chapter 10 for the stress tensor demonstrates that these principal values are stationary values, including the largest and smallest values, of $\mathbf{n} \cdot \mathbf{D} \cdot \mathbf{n}$ over all orientations.

To investigate the meaning of the off-diagonal components of $\mathbf{D}$ we consider the rate of change of the scalar product between two infinitesimal line segments $d\mathbf{x}_A$ and $d\mathbf{x}_B$

$$d\mathbf{x}_A \cdot d\mathbf{x}_B = ds_A\,ds_B\cos\theta$$

where ds_A and ds_B are the lengths of $d\mathbf{x}_A$ and $d\mathbf{x}_B$, respectively, and θ is the angle between them. Taking the time derivative of both sides,

$$\frac{d}{dt}(d\mathbf{x}_A \cdot d\mathbf{x}_B) = \frac{d}{dt}(ds_A ds_B \cos\theta)$$

gives

$$d\mathbf{v}_A \cdot d\mathbf{x}_B + d\mathbf{x}_A \cdot d\mathbf{v}_B = \frac{d}{dt}(ds_A)\,ds_B\cos\theta + ds_A\frac{d}{dt}(ds_B)\cos\theta - ds_A\,ds_B\sin\theta\dot{\theta} \qquad (14.6)$$

Using (14.2) and regrouping yields

$$2\frac{d\mathbf{x}_A}{ds_A} \cdot \mathbf{D} \cdot \frac{d\mathbf{x}_B}{ds_B} = \left\{\frac{1}{ds_A}\frac{d}{dt}(ds_A) + \frac{1}{ds_B}\frac{d}{dt}(ds_B)\right\}\cos\theta - \sin\theta\dot{\theta} \qquad (14.7)$$

Note that because $d\mathbf{x}_A$ and $d\mathbf{x}_B$ are infinitesimal line segments emanating from the same point, the same value of $\mathbf{L}$ is used for each. When $\theta = 90°$, the line segments are orthogonal and (14.7) reduces to

$$\mathbf{n}_A \cdot \mathbf{D} \cdot \mathbf{n}_B = -\frac{1}{2}\dot{\theta}$$

Thus, the off-diagonal components give half the rate of decrease of the angle between linear segments aligned with the coordinate directions.

14.3 Meaning of W

Since $\mathbf{W}$ is an anti- (or skew-) symmetric tensor, $\mathbf{W} = -\mathbf{W}^T$, it has only three distinct nonzero components. These components can be associated with a vector $\mathbf{w}$ by means of the following operation:

$$\mathbf{W} \cdot \mathbf{a} = \mathbf{w} \times \mathbf{a} \tag{14.8}$$

where $\mathbf{a}$ is an arbitrary vector. The vector $\mathbf{w}$ is called the dual or polar vector (of a skew-symmetric tensor). Writing (14.8) in component form and recognizing that this relation must apply for any vector $\mathbf{a}$ yields the component form of $\mathbf{W}$:

$$W_{ij} = \epsilon_{imj} w_m$$

Multiplying both sides with ϵ_{ijp} and using the ϵ–δ identity (4.13) yields

$$w_q = -\frac{1}{2}\epsilon_{qip} W_{ip} \tag{14.9}$$

The polar vector can be related to the velocity field by substituting the component form of $\mathbf{W}$ into (14.9)

$$w_i = -\frac{1}{2}\epsilon_{ijk}\frac{1}{2}\left(\frac{\partial v_j}{\partial x_k} - \frac{\partial v_k}{\partial x_j}\right) = \frac{1}{2}\epsilon_{ijk}\partial_j v_k$$

or

$$\mathbf{w} = \frac{1}{2}(\nabla \times v)$$

The combination $\nabla \times v$ is called the *vorticity*. If $\mathbf{w} = 0$, so is the vorticity, and the velocity field is said to be *irrotational*. Because

$$\nabla \times \nabla \phi = 0$$

for any scalar field ϕ, in an irrotational field the velocity vector can be represented as the gradient of a scalar, i.e., $\mathbf{v} = \nabla \phi$.

Now, suppose $\mathbf{D} = 0$:

$$d\mathbf{v} = \mathbf{W} \cdot d\mathbf{x}$$

Using (14.8) gives

$$d\mathbf{v} = \mathbf{w} \times d\mathbf{x}$$

Hence, the local velocity increment is a rigid spin with angular velocity $\mathbf{w}$.

Exercises

14.1   Show that $\lambda = 0$ is the *only real* principal value of an antisymmetric tensor, $\mathbf{W} = -\mathbf{W}^T$.

14.2   Show that the axial vector of an antisymmetric tensor of the form $\mathbf{W} = \mathbf{ba} - \mathbf{ab}$ is $\mathbf{a} \times \mathbf{b}$.

14.3   **(a)**   Let $\mathbf{p}$ be a unit vector in the principal direction of $\mathbf{W}$ corresponding to $\lambda = 0$ and, hence, satisfying

$$\mathbf{W} \cdot \mathbf{p} = 0$$

Explain why $\mathbf{p}$ is parallel to $\mathbf{w}$ and, consequently, $\mathbf{w} = w\mathbf{p}$, where $\mathbf{w}$ is the *axial vector* or *polar vector* of $\mathbf{W}$ defined by (14.8).

(b)   Let $\mathbf{q}$ and $\mathbf{r}$ be unit vectors orthogonal to each other ($\mathbf{q} \cdot \mathbf{r} = 0$) and to $\mathbf{p}$ ($\mathbf{q} \cdot \mathbf{p} = 0$, $\mathbf{r} \cdot \mathbf{p} = 0$) such that $\mathbf{p}$, $\mathbf{q}$, and $\mathbf{r}$ form a right-handed system, $\mathbf{p} \cdot \mathbf{q} \times \mathbf{r} = 1$. Show that $\mathbf{W}$ is given by

$$\mathbf{W} = w(\mathbf{rq} - \mathbf{qr})$$

where $w = \mathbf{r} \cdot \mathbf{W} \cdot \mathbf{q}$. (Problem 4.5.a may be useful here.)

14.4   Consider the motion

$$\mathbf{x}(\mathbf{X}, t) = \mathbf{Q}(t) \cdot \mathbf{X} + \mathbf{c}(t)$$

(a)   Determine expressions for the rate-of-deformation and spin tensors.
(b)   If the rate-of-deformation tensor vanishes, the motion is rigid and the lengths of lines remain the same in the current and reference configurations. In this case show that $\mathbf{Q}(t)$ is an orthogonal tensor, i.e., $\mathbf{Q}^{-1}(t) = \mathbf{Q}^T(t)$, and that

$$\mathbf{W} = \dot{\mathbf{Q}}(t) \cdot \mathbf{Q}^{-1}(t)$$

is antisymmetric.
(c)   Show that for a rigid motion (see Example 13.1) the Eulerian descriptions of velocity and acceleration can be written as

$$\mathbf{v}(\mathbf{x}, t) = \dot{\mathbf{c}}(t) + \mathbf{W} \cdot (\mathbf{x} - \mathbf{c})$$
$$\mathbf{a}(\mathbf{x}, t) = \ddot{\mathbf{c}}(t) + (\dot{\mathbf{W}} + \mathbf{W} \cdot \mathbf{W}) \cdot (\mathbf{x} - \mathbf{c})$$

(d)   Or, in terms of the axial vector of $\mathbf{W}$,

$$\mathbf{v}(\mathbf{x}, t) = \dot{\mathbf{c}}(t) + \mathbf{w} \times (\mathbf{x} - \mathbf{c})$$
$$\mathbf{a}(\mathbf{x}, t) = \ddot{\mathbf{c}}(t) + \dot{\mathbf{w}} \times (\mathbf{x} - \mathbf{c}) + \mathbf{w} \times \{\mathbf{w} \times (\mathbf{x} - \mathbf{c})\}$$

14.5 In rigid motion of a body, the spatial positions of four particles relative to a fixed origin are given by the vectors $\mathbf{c}$ and $\mathbf{c} + \mathbf{a}_i$, where the $\mathbf{a}_i$ are orthonormal vectors. Show that the spin tensor $\mathbf{W}$ and the polar vector $\mathbf{w}$ are given by

$$\mathbf{W} = \frac{1}{2}(\dot{\mathbf{a}}_p \mathbf{a}_p - \mathbf{a}_p \dot{\mathbf{a}}_p)$$

$$\mathbf{w} = \frac{1}{2}\mathbf{a}_p \times \dot{\mathbf{a}}_p$$

15

Geometric Measures of Deformation

In the preceding chapter we were concerned only with the instantaneous rate of deformation and spin in the current configuration. Now we want to compare the geometry in the current configuration to that in the reference configuration.

15.1 Deformation Gradient

Figure 15.1 shows an infinitesimal line segment in the reference configuration, $d\mathbf{X}$, mapped into an infinitesimal line segment in the current configuration, $d\mathbf{x}$, by

$$dx_k = \frac{\partial x_k}{\partial X_m} dX_m$$

where $\partial x_k / \partial X_m$ are components of the *deformation gradient tensor*:

$$F_{km} = \frac{\partial x_k}{\partial X_m}(\mathbf{X}) \tag{15.1}$$

Note that in $\mathbf{F}$ the gradient is with respect to position in the reference configuration. In coordinate-free notation

$$d\mathbf{x} = \mathbf{F} \cdot d\mathbf{X} = d\mathbf{X} \cdot \mathbf{F}^T \tag{15.2}$$

We will show that the tensor $\mathbf{F}$ contains all information about the geometry of deformation: change in length of lines, change in angles, change in area, and change in volume.

Fundamentals of Continuum Mechanics, First Edition. John W. Rudnicki.
© 2015 John Wiley & Sons, Ltd. Published 2015 by John Wiley & Sons, Ltd.

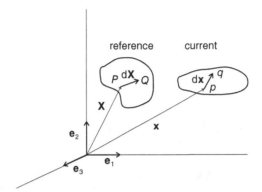

Figure 15.1   Infinitesimal line segments in the reference and current configurations.

15.2   Change in Length of Lines

The square of the length of an infinitesimal line segment $d\mathbf{x}$ in the current configuration is given by its scalar product with itself:

$$(ds)^2 = d\mathbf{x} \cdot d\mathbf{x} = (d\mathbf{X} \cdot \mathbf{F}^T) \cdot (\mathbf{F} \cdot d\mathbf{X}) \tag{15.3}$$

$$= d\mathbf{X} \cdot (\mathbf{F}^T \cdot \mathbf{F}) \cdot d\mathbf{X} \tag{15.4}$$

The length of the line segment in the reference configuration is $dS = (d\mathbf{X} \cdot d\mathbf{X})^{1/2}$ and $\mathbf{N} = d\mathbf{X}/dS$ is a unit vector in the direction of the infinitesimal line segment $d\mathbf{X}$ in the reference configuration. Now (15.4) can be written as

$$\left(\frac{ds}{dS}\right)^2 = \mathbf{N} \cdot (\mathbf{F}^T \cdot \mathbf{F}) \cdot \mathbf{N}$$

The ratio

$$\frac{ds}{dS} = \Lambda(\mathbf{N}) \tag{15.5}$$

defines the *stretch ratio*. The tensor

$$\mathbf{C} = \mathbf{F}^T \cdot \mathbf{F} \tag{15.6}$$

is called the *Green deformation tensor* by Malvern (1988) or the *right Cauchy–Green tensor* by Truesdell and Noll (1965). Note that $\mathbf{C}$ is symmetric. (See Problem 15.1.)

The stretch ratio (15.5) can be expressed as

$$\Lambda = \frac{ds}{dS} = \sqrt{\mathbf{N} \cdot \mathbf{C} \cdot \mathbf{N}} \tag{15.7}$$

Because $\mathbf{C}$ is symmetric, it possesses three real positive principal values that can be associated with three orthogonal principal directions. The principal values are *squares* of the principal stretch ratios, Λ_I, Λ_{II}, Λ_{III}, with corresponding principal directions $\mathbf{N}_I$, $\mathbf{N}_{II}$, $\mathbf{N}_{III}$. Because the stretch ratios must be positive, $\mathbf{C}$ is positive definite, i.e.,

$$\mathbf{x} \cdot \mathbf{C} \cdot \mathbf{x} > 0$$

for any vector $\mathbf{x} \neq 0$. By the same derivation as in Chapter 10, the principal stretch ratios include the largest and smallest values of the stretch ratio. Thus, $\mathbf{C}$ has the following principal axes representation in dyadic form:

$$\mathbf{C} = \Lambda_I^2 \mathbf{N}_I \mathbf{N}_I + \Lambda_{II}^2 \mathbf{N}_{II} \mathbf{N}_{II} + \Lambda_{III}^2 \mathbf{N}_{III} \mathbf{N}_{III}$$

Because each line segment in the current configuration must originate from a line segment in the reference configuration, the tensor $\mathbf{F}$ possesses an inverse:

$$d\mathbf{X} = \mathbf{F}^{-1} \cdot d\mathbf{x} = d\mathbf{x} \cdot \mathbf{F}^{-1T}$$

Consequently, it is possible to calculate the reciprocal of the ratio (15.5), $\lambda = \Lambda^{-1}$, in terms of $\mathbf{F}^{-1}$:

$$(dS)^2 = d\mathbf{X} \cdot d\mathbf{X} = d\mathbf{x} \cdot (\mathbf{F}^{-1T} \cdot \mathbf{F}^{-1}) \cdot d\mathbf{x}$$

or

$$\lambda^2 = \mathbf{n} \cdot (\mathbf{F}^{-1T} \cdot \mathbf{F}^{-1}) \cdot \mathbf{n} \tag{15.8}$$

where $\mathbf{n} = d\mathbf{x}/ds$ is a unit vector in the direction of the line segment in the current configuration. The inverse of the tensor

$$\mathbf{B} = \mathbf{F} \cdot \mathbf{F}^T$$

is equal to the product in parentheses on the right side of (15.8). The tensor $\mathbf{B}$ is called the *left Cauchy–Green tensor* by Truesdell and Noll (1965). Its inverse $\mathbf{B}^{-1}$ (sometimes denoted $\mathbf{c}$) is called the *Cauchy deformation tensor* by Malvern (1988). Prager (1973) calls $\mathbf{B}$ and $\mathbf{B}^{-1}$ the Finger tensors.

15.3   Change in Angles

The angle θ between two line segments $d\mathbf{x}_A$ and $d\mathbf{x}_B$ in the current configuration (Figure 15.2) is given by

$$\cos\theta = \frac{d\mathbf{x}_A \cdot d\mathbf{x}_B}{|d\mathbf{x}_A| \, |d\mathbf{x}_B|}$$

Using (15.2) and (15.6) yields

$$\cos\theta = \frac{\mathbf{N}_A \cdot \mathbf{C} \cdot \mathbf{N}_B}{(\mathbf{N}_A \cdot \mathbf{C} \cdot \mathbf{N}_A)^{1/2}(\mathbf{N}_B \cdot \mathbf{C} \cdot \mathbf{N}_B)^{1/2}} \tag{15.9}$$

Because $d\mathbf{X}_A$ and $d\mathbf{X}_B$ are infinitesimal line segments emanating from the same point, the deformation gradient $\mathbf{F}$ is the same for both. The terms in the denominator of (15.9) are the

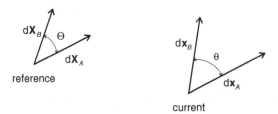

reference

current

Figure 15.2 Angle Θ between two infinitesimal line segments in the reference configuration changes to θ in the current configuration.

stretch ratios in the directions $\mathbf{N}_A$ and $\mathbf{N}_B$, Λ_A and Λ_B, respectively. We define the *shear* as the change in angle between line segments in the directions $\mathbf{N}_A$, $\mathbf{N}_B$ in the reference configuration:

$$\gamma(\mathbf{N}_A, \mathbf{N}_B) = \Theta - \theta$$

where

$$\cos\Theta = \mathbf{N}_A \cdot \mathbf{N}_B$$

Using (15.9) gives

$$\cos(\Theta - \gamma) = \mathbf{N}_A \cdot \mathbf{C} \cdot \mathbf{N}_B \left(\Lambda_A \Lambda_B\right)^{-1} \tag{15.10}$$

In the special case, $\Theta = 90°$, $\cos(90° - \gamma) = \sin\gamma$. Note that if $\mathbf{N}_A$ and $\mathbf{N}_B$ are principal directions, $\gamma = 0$ (because $\mathbf{N}_A \cdot \mathbf{C} \cdot \mathbf{N}_B = \Lambda_B^2 \mathbf{N}_B \cdot \mathbf{N}_A = 0$). Therefore principal directions of $\mathbf{C}$ in the reference configuration remain orthogonal in the current configuration.

15.4   Change in Area

An oriented element of area in the reference configuration is given by

$$\mathbf{N}\,dA = d\mathbf{X}_A \times d\mathbf{X}_B$$
$$= \mathbf{e}_i \epsilon_{ijk}(d\mathbf{X}_A)_j (d\mathbf{X}_B)_k$$

and is deformed into

$$\mathbf{n}\,da = d\mathbf{x}_A \times d\mathbf{x}_B \tag{15.11}$$

in the current configuration. Substituting (15.2) into the component form of (15.11) yields

$$n_i\,da = \epsilon_{ijk} \left[F_{jr}(dX_A)_r\right] \left[F_{ks}(dX_B)_s\right]$$

Multiplying both sides by F_{it} gives

$$n_i F_{it}\,da = \left[\epsilon_{ijk} F_{it} F_{jr} F_{ks}\right] (dX_A)_r (dX_B)_s$$

The term in square brackets can be written in terms of the determinant of $\mathbf{F}$ (5.3). The result is

$$n_i F_{it}\, da = \epsilon_{rst}\, \det(\mathbf{F})(dX_A)_r (dX_B)_s$$

Reverting to coordinate-free notation gives

$$\mathbf{n} \cdot \mathbf{F}\, da = \det(\mathbf{F})\, d\mathbf{X}_A \times d\mathbf{X}_B$$
$$= \det(\mathbf{F})\mathbf{N}\, dA$$

and then multiplying (from the right) by $\mathbf{F}^{-1}$ gives *Nanson's formula* relating areas in the current and reference configurations:

$$\mathbf{n}\, da = \det(\mathbf{F})(\mathbf{N} \cdot \mathbf{F}^{-1})\, dA \tag{15.12}$$

15.5   Change in Volume

An element of volume in the reference configuration is given by the triple scalar product of three line segments $d\mathbf{X}_A$, $d\mathbf{X}_B$, and $d\mathbf{X}_C$

$$dV = d\mathbf{X}_A \cdot (d\mathbf{X}_B \times d\mathbf{X}_C) = \epsilon_{ijk}(dX_A)_i(dX_B)_j(dX_C)_k$$

Similarly, an element of volume in the current configuration is

$$dv = d\mathbf{x}_A \cdot (d\mathbf{x}_B \times d\mathbf{x}_C) = \epsilon_{rst}(dx_A)_r(dx_B)_s(dx_C)_t$$

Substituting (15.2) and rearranging gives

$$dv = \epsilon_{rst} F_{ri} F_{sj} F_{tk}(dX_A)_i(dX_B)_j(dX_C)_k$$

Again using (5.3) gives

$$dv = \det(\mathbf{F})\epsilon_{ijk}(dX_A)_i(dX_B)_j(dX_C)_k$$

and reverting to coordinate-free notation gives

$$dv = \det(\mathbf{F})d\mathbf{X}_A \cdot (d\mathbf{X}_B \times d\mathbf{X}_C) \tag{15.13}$$

Defining J as the ratio of current to reference volume elements gives

$$J = \frac{dv}{dV} = \frac{\rho_0}{\rho} = \det(\mathbf{F}) \tag{15.14}$$

Because the ratios dv/dV and ρ_0/ρ must be strictly positive, $\det(\mathbf{F}) > 0$, establishing that $\mathbf{F}^{-1}$ does exist. That $\det(\mathbf{F}) > 0$ is identical to the condition (13.5) confirms the earlier assertion that the motion can be inverted (13.2).

Because of (15.6) and the result of Example 5.3

$$\det(\mathbf{C}) = \det \mathbf{F}^T \det \mathbf{F} = (\det \mathbf{F})^2$$

and (15.13) and (15.14) can be expressed in terms of $\mathbf{C}$. Similarly, changes in the length of lines and angles are expressed in terms of $\mathbf{C}$ (rather than $\mathbf{F}$ alone). Hence, it is $\mathbf{C}$ that describes deformation. The expression for the change in area (15.12) involves $\mathbf{F}$ because a change in area can occur by pure rotation without deformation.

15.6   Polar Decomposition

In the discussion of shear and angle change following (15.10), we noted that a triad in the directions of principal stretch ratios remains orthogonal after deformation. That is, the shear is zero for two lines in the principal directions of $\mathbf{C}$ in the reference configuration. Consequently, we can imagine the deformation to occur in the two steps shown schematically in Figure 15.3: First, a pure deformation that stretches line elements in the principal directions to their final length; then a rotation that orients these line elements in the proper directions in the current configuration.

The deformation is given by

$$d\mathbf{x}' = \mathbf{U} \cdot d\mathbf{X} \tag{15.15}$$

Because $\mathbf{U}$ is the deformation tensor that stretches line elements in the principal directions, it has the same principal directions as $\mathbf{C}$ and has principal values that are equal to the principal stretch ratios. Hence, the principal axis representation of $\mathbf{U}$ is

$$\mathbf{U} = \Lambda_I \mathbf{N}_I \mathbf{N}_I + \Lambda_{II} \mathbf{N}_{II} \mathbf{N}_{II} + \Lambda_{III} \mathbf{N}_{III} \mathbf{N}_{III}$$

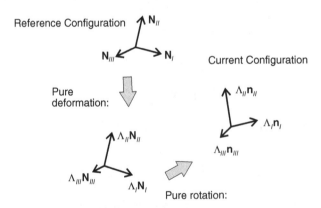

Figure 15.3   Illustration of the polar decomposition of deformation into a pure stretching and a pure rotation.

where $\mathbf{U} = \mathbf{U}^T$. Note that although $\mathbf{U}$ does not change the right angle between principal directions, in general the angles between lines that are not oriented in the principal directions will change.

Then the principal directions in the reference configuration are rotated into their proper orientation in the current configuration:

$$d\mathbf{x} = \mathbf{R} \cdot d\mathbf{x}' \qquad (15.16)$$

$\mathbf{R}$ is an orthogonal tensor corresponding to a pure rotation, so that the lengths of line segments will be preserved (Chapter 2). Combining (15.15) and (15.16) yields

$$d\mathbf{x} = \mathbf{F} \cdot d\mathbf{X} = (\mathbf{R} \cdot \mathbf{U}) \cdot d\mathbf{X}$$

Thus, the deformation gradient tensor is decomposed into the product of a pure deformation tensor and a rotation tensor:

$$\mathbf{F} = \mathbf{R} \cdot \mathbf{U} \qquad (15.17)$$

Substituting (15.17) into the expression for the Green deformation tensor (15.6) gives

$$\mathbf{C} = \mathbf{F}^T \cdot \mathbf{F} = \mathbf{U}^T \cdot \mathbf{U} = \mathbf{U}^2$$

Formally, we can write $\mathbf{U} = \sqrt{\mathbf{C}}$, but this operation can be carried out only in principal axis form. In order to calculate the components of $\mathbf{U}$ from $\mathbf{C}$ it is necessary to express $\mathbf{C}$ in principal axis form, take the square roots of the principal values, then convert back to the coordinate system of interest.

Alternatively, we could have rotated first, then stretched. This leads to

$$d\mathbf{x} = \mathbf{V} \cdot \mathbf{R} \cdot d\mathbf{X}$$

where

$$\mathbf{V} = \lambda_I^{-1}\mathbf{n}_I\mathbf{n}_I + \lambda_{II}^{-1}\mathbf{n}_{II}\mathbf{n}_{II} + \lambda_{III}^{-1}\mathbf{n}_{III}\mathbf{n}_{III}$$

and

$$\mathbf{n}_K = \mathbf{R} \cdot \mathbf{N}_K$$

The rotation tensor is given by the dyad

$$\mathbf{R} = \mathbf{n}_K\mathbf{N}_K$$

and $\mathbf{U}$ and $\mathbf{V}$ are related by

$$\mathbf{V} = \mathbf{R} \cdot \mathbf{U} \cdot \mathbf{R}^T$$

Thus, $\mathbf{U}$ and $\mathbf{V}$ have the same principal values but their principal directions are related by the rotation tensor $\mathbf{R}$.

15.7   Example

Show that an infinitesimal sphere in the reference configuration

$$d\mathbf{X} \cdot d\mathbf{X} = 1 \tag{15.18}$$

deforms into an ellipsoid in the current configuration with semi-axes given by the principal stretch ratios and with axes aligned with principal directions in the current configuration.

We begin with

$$d\mathbf{X} = \mathbf{F}^{-1} \cdot d\mathbf{x} = (\mathbf{V} \cdot \mathbf{R})^{-1} \cdot d\mathbf{x} = \left(\mathbf{R}^T \cdot \mathbf{V}^{-1} \right) \cdot d\mathbf{x} \tag{15.19}$$

where we have used $\mathbf{R}^{-1} = \mathbf{R}^T$. Similarly,

$$d\mathbf{X} = d\mathbf{x} \cdot \left(\mathbf{F}^{-1} \right)^T = d\mathbf{x} \cdot \left(\mathbf{R}^T \cdot \mathbf{V}^{-1} \right)^T = d\mathbf{x} \cdot \left(\mathbf{V}^{-1T} \cdot \mathbf{R} \right) \tag{15.20}$$

Substituting (15.19) and (15.20) into (15.18) yields

$$d\mathbf{x} \cdot \left(\mathbf{V}^{-1T} \cdot \mathbf{V}^{-1} \right) \cdot d\mathbf{x} = 1 \tag{15.21}$$

Writing $\mathbf{V}^{-1}$ in principal axis form

$$\mathbf{V}^{-1} = \sum_{K=I,II,III} \lambda_K \mathbf{n}_K \mathbf{n}_K$$

and substituting into (15.21) yields

$$\left(\lambda_I \, d\mathbf{x} \cdot \mathbf{n}_I \right)^2 + \left(\lambda_{II} \, d\mathbf{x} \cdot \mathbf{n}_{II} \right)^2 + \left(\lambda_{III} \, d\mathbf{x} \cdot \mathbf{n}_{III} \right)^2 = 1$$

Exercises

15.1   Show that $\mathbf{C}$ is symmetric.

15.2   Show that $\det \mathbf{F} = \det \mathbf{U}$.

15.3   The motion of a continuum is given by

$$x_1 = \alpha \cos(\theta) X_1 + \beta \sin(\theta) X_2$$
$$x_2 = -\alpha \sin(\theta) X_1 + \beta \cos(\theta) X_2$$
$$x_3 = X_3$$

where α, β, and θ are constants.
(a)   Determine the deformation tensor $\mathbf{F}$.
(b)   Determine the deformation tensor $\mathbf{C}$.
(c)   Determine the deformation tensor $\mathbf{U}$.
(d)   Determine the rotation tensor $\mathbf{R}$.
(e)   Determine the principal directions of $\mathbf{V}$.

15.4 The definition of the velocity gradient tensor $\mathbf{L}$ involved the gradient with respect to the current coordinates $[\mathbf{\nabla}_{\mathbf{x}}(\ldots)]_i = \partial(\ldots)/\partial x_i$, whereas the deformation gradient tensor involves gradients with respect to the reference coordinates $[\mathbf{\nabla}_{\mathbf{X}}(\ldots)]_i = \partial(\ldots)/\partial X_i$. Show that for a position-dependent vector $\mathbf{u}$, these gradients are related by

$$\mathbf{\nabla}_{\mathbf{X}}\mathbf{u} = \mathbf{F}^T \cdot (\mathbf{\nabla}_{\mathbf{x}}\mathbf{u})$$

15.5 Show that the magnitudes of the areas in the current and reference configurations are related by

$$\frac{\mathrm{d}a}{\mathrm{d}A} = \det(\mathbf{F})\sqrt{\mathbf{N} \cdot \mathbf{C}^{-1} \cdot \mathbf{N}}$$

15.6 Show that the angle Θ between two lines in the reference configuration that are in the directions $\mathbf{n}_A$ and $\mathbf{n}_B$ in the current configuration is given by

$$\cos\Theta = \lambda_A^{-1}\lambda_B^{-1}\mathbf{n}_A \cdot \mathbf{B}^{-1} \cdot \mathbf{n}_B$$

where λ_A and λ_B are the reciprocals of stretch ratios in directions $\mathbf{n}_A$ and $\mathbf{n}_B$ in the current configuration.

15.7 Show that the ratio $(\mathrm{d}A/\mathrm{d}a)^2$ is given by

$$\left(\frac{\mathrm{d}A}{\mathrm{d}a}\right)^2 = J^{-2}\mathbf{n} \cdot \mathbf{B} \cdot \mathbf{n}$$

where the area element with magnitude $\mathrm{d}a$ has the unit normal $\mathbf{n}$.

15.8 Show that the deformation tensor $\mathbf{U}$ is symmetric in any rectangular Cartesian coordinate system (and, hence, in any coordinate system).

15.9 Show that

$$\Lambda\mathbf{n} = \mathbf{F} \cdot \mathbf{N}$$

15.10 A simple shear deformation deforms a unit square into a rhombus as shown in Figure 13.4 and is described by

$$x_1 = X_1 + \gamma X_2, \quad x_2 = X_2, \quad x_3 = X_3$$

Determine the deformation gradient tensor $\mathbf{F}$ and the Cauchy deformation tensor $\mathbf{C} = \mathbf{F}^T \cdot \mathbf{F}$.

15.11 For the simple shear deformation of Problem 15.10 use the formula (15.7) to compute the stretch ratios of both diagonals and verify your results using geometry.

15.12 For the simple shear deformation of Problem 15.10 with $\gamma = 1$:
   (a) Compute the principal values of $\mathbf{C}$.
   (b) Compute the principal stretches. [Answer: $\Lambda_I = 1.681$]
   (c) Compute the angle that the principal direction corresponding to the largest principal value of $\mathbf{C}$ makes with $\mathbf{e}_1$. [Answer: $58.3°$]

15.13 For the simple shear deformation:

    (a) Show that the principal stretch ratios (in the reference state) can be expressed as $\Lambda_I = \alpha = \Lambda_{II}{}^{-1}$ and $\Lambda_{III} = 1$ where

$$\alpha = \sqrt{1 + (\gamma/2)^2} + (\gamma/2)$$

    (b) Show that the corresponding principal directions are given by

$$\mathbf{N}_K = \mathbf{R}^L \cdot \mathbf{e}_k$$

where

$$\mathbf{R}^L = \cos(\Theta)(\mathbf{e}_1\mathbf{e}_1 + \mathbf{e}_2\mathbf{e}_2) + \sin(\Theta)(\mathbf{e}_2\mathbf{e}_1 - \mathbf{e}_1\mathbf{e}_2) + \mathbf{e}_3\mathbf{e}_3 \qquad (15.22)$$

and $\tan(\Theta) = \alpha$.

    (c) Show that the components of the deformation tensor $\mathbf{U}$ are given by

$$U_{11} = \alpha\cos^2(\Theta) + \alpha^{-1}\sin^2(\Theta) = \frac{2\alpha}{\alpha^2 + 1}$$

$$U_{12} = U_{21} = \left(\alpha - \alpha^{-1}\right)\cos(\Theta)\sin(\Theta) = \frac{\alpha^2 - 1}{\alpha^2 + 1}$$

$$U_{22} = \alpha^{-1}\cos^2(\Theta) + \alpha\sin^2(\Theta) = \frac{\alpha^3 + \alpha^{-1}}{\alpha^2 + 1}$$

$$U_{33} = 1$$

15.14 For the simple shear deformation:

    (a) Show that the rotation tensor $\mathbf{R}$ is given by

$$\mathbf{R} = \cos(\omega)(\mathbf{e}_1\mathbf{e}_1 + \mathbf{e}_2\mathbf{e}_2) - \sin(\omega)(\mathbf{e}_2\mathbf{e}_1 - \mathbf{e}_1\mathbf{e}_2) + \mathbf{e}_3\mathbf{e}_3$$

where $\tan\omega = \gamma/2$. Thus, the principal directions in the current state are given by

$$\mathbf{n}_K = \mathbf{R}^E \cdot \mathbf{e}_k$$

where $\mathbf{R}^E$ has the same form as in (15.22) with Θ replaced by $\theta = \Theta - \omega$.

    (b) Show that $\Theta + \theta = \pi/2$.

References

Malvern LE 1988 *Introduction to the Mechanics of a Continuous Medium*. Prentice Hall.

Prager W 1973 *Introduction to the Mechanics of Continua*. Dover.

Truesdell C and Noll W 1965 The non-linear field theories of mechanics. In *Encyclopedia of Physics* (ed. Flügge S) Springer-Verlag (3rd edition, 2004, ed. Antman S).

16

Strain Tensors

16.1 Material Strain Tensors

For an appropriate material strain tensor, we want it to characterize line length, angle, and volume changes but to be independent of any rigid rotation. For finite strain, there are many possibilities but it is sensible that they all agree with the small-strain tensor when strains are indeed small. Thus, a material strain tensor is defined by the following requirements (Hill 1968):

1. Has the same principal axes as $\mathbf{U}$.
2. Vanishes when the principal stretch ratios are unity.
3. Agrees with the small-strain tensor.

The first requirement constrains a material strain tensor to have the following principal axis form:

$$\mathbf{E} = f(\Lambda_I)\mathbf{N}_I\mathbf{N}_I + f(\Lambda_{II})\mathbf{N}_{II}\mathbf{N}_{II} + f(\Lambda_{III})\mathbf{N}_{III}\mathbf{N}_{III} \qquad (16.1)$$

where the $\mathbf{N}_K$ are the principal directions of $\mathbf{U}$, the Λ_K are the principal stretch ratios (the square root of the principal values of $\mathbf{C}$), and $f(\Lambda)$ is a smooth and monotonic, but otherwise arbitrary, function. By construction, a material strain tensor is symmetric, $\mathbf{E} = \mathbf{E}^T$. The second requirement restricts the value of $f(1) = 0$ so that $\mathbf{E} = \mathbf{0}$ when $\mathbf{U} = \mathbf{I}$. The last requires $f'(1) = 1$ so that $\mathbf{E}$ agrees with the small-strain tensor. To demonstrate this, we expand $f(\Lambda)$ about $\Lambda = 1$:

$$f(\Lambda) = f(1) + f'(1)(\Lambda - 1) + \frac{1}{2}f''(1)(\Lambda - 1)^2 + \dots$$

Using $f(1) = 0, f'(1) = 1$, and retaining only the linear term yields

$$f(\Lambda) = \Lambda - 1$$

Fundamentals of Continuum Mechanics, First Edition. John W. Rudnicki.
© 2015 John Wiley & Sons, Ltd. Published 2015 by John Wiley & Sons, Ltd.

Thus, the principal values of $\mathbf{E}$ reduce to change in length per unit (reference) length for principal stretch ratios near unity.

The most common choice for the scale function $f(\Lambda)$ is

$$f(\Lambda) = \frac{1}{2}(\Lambda^2 - 1) \tag{16.2}$$

Substituting (16.2) into (16.1) and combining terms defines the *Green (Lagrangian)* strain tensor

$$\mathbf{E}^G = \frac{1}{2}(\mathbf{U}^T \cdot \mathbf{U} - \mathbf{I})$$

This is a convenient choice because $\mathbf{U}^2 = \mathbf{U}^T \cdot \mathbf{U}$ can be calculated directly from the deformation tensor $\mathbf{F}$:

$$\mathbf{E}^G = \frac{1}{2}(\mathbf{F}^T \cdot \mathbf{F} - \mathbf{I}) \tag{16.3}$$

Determining $\mathbf{U}$ (or any odd power of $\mathbf{U}$) requires first finding the principal values and directions of $\mathbf{C}$. For arbitrary stretch ratios, the normal components of the Green–Lagrange strain do not give change in length per unit reference length but, as indicated by (16.2), the current length squared minus the reference length squared divided by two times the reference length squared.

The component form of (16.3) is

$$E_{ij}^G = \frac{1}{2}\left(F_{ik}^T F_{kj} - \delta_{ij}\right) = \frac{1}{2}(F_{ki}F_{kj} - \delta_{ij}) \tag{16.4}$$

or using (15.1), $F_{kl} = \partial x_k / \partial X_l$, gives

$$E_{ij}^G = \frac{1}{2}\left\{ \frac{\partial x_k}{\partial X_i} \frac{\partial x_k}{\partial X_j} - \delta_{ij} \right\} \tag{16.5}$$

E_{ij}^G can be expressed in terms of the displacement components u_k by noting that $x_k = X_k + u_k$:

$$E_{ij}^G = \frac{1}{2}\left\{ \frac{\partial u_i}{\partial X_j} + \frac{\partial u_j}{\partial X_i} + \frac{\partial u_k}{\partial X_j} \frac{\partial u_k}{\partial X_i} \right\} \tag{16.6}$$

Although the choice of (16.2) is the most common one for the scale function, there are many other possibilities. Perhaps the most obvious extension of small strain is to choose

$$f(\Lambda) = \Lambda - 1 = \frac{\text{change in length}}{\text{reference length}}$$

Using (16.1) to convert to tensor form yields

$$\mathbf{E}^{(1)} = \mathbf{U} - \mathbf{I} \tag{16.7}$$

This is a finite strain measure that was introduced and used by Biot (1965), but has the drawback that it cannot be expressed directly in terms of $\mathbf{F}$. Another possibility corresponds to defining normal strains as change in length per unit *current* length:

$$\mathbf{E}^{(-1)} = \mathbf{I} - \mathbf{U}^{-1}$$

Still another possibility is logarithmic strain, corresponding to the scale function $f(\Lambda) = \ln \Lambda$. This is often used as the large-strain measure for uniaxial bar tests. This one-dimensional measure can be extended to a tensor version in a manner similar to other finite strains:

$$\mathbf{E}^{(\ln)} = \ln \mathbf{U}$$

Thus $\mathbf{E}^{(\ln)}$ has the same principal directions as $\mathbf{U}$ but principal values that are the logarithms of the principal stretch ratios. The operation implied by the right side of only makes sense in principal axes form; for axes that are not aligned with the principal directions, the components of $\mathbf{E}^{(\ln)}$ are not the logarithms of the components of $\mathbf{U}$.

A form of the scale function that includes all of these strain measures as special cases is (Ogden 1997)

$$f(\Lambda) = \frac{1}{m}(\Lambda^m - 1) \tag{16.8}$$

If m is even, the strain can be written directly in terms of the deformation gradient. The Green–Lagrange strain (16.3) corresponds to $m = 2$, and (16.7) to $m = 1$. The limit $m \to 0$ yields the logarithmic strain measure.

16.2   Spatial Strain Measures

We can define a class of spatial strain measures in a manner analogous to the material strain measures. The spatial strain measures have the same principal axes as $\mathbf{V}$

$$\mathbf{e} = g(\lambda_I)\mathbf{n}_I\mathbf{n}_I + g(\lambda_{II})\mathbf{n}_{II}\mathbf{n}_{II} + g(\lambda_{III})\mathbf{n}_{III}\mathbf{n}_{III} \tag{16.9}$$

and the same requirements on the scale functions: $g(1) = 1$ and $g'(1) = -1$. In (16.9) we use $\lambda = \Lambda^{-1} = dS/ds$ to emphasize that we are working in the current configuration.

The most commonly used spatial strain measure is the *Almansi strain* corresponding to the scale function

$$g(\lambda) = \frac{1}{2}(1 - \lambda^2)$$

Converting to tensor form yields

$$\mathbf{e}^A = \frac{1}{2}\left\{\mathbf{I} - \mathbf{F}^{(-1)T} \cdot \mathbf{F}^{-1}\right\} \tag{16.10}$$

Expressing this in terms of Cartesian components gives

$$e^A_{ij} = \frac{1}{2}\left\{\delta_{ij} - F^{-1}_{ki} F^{-1}_{kj}\right\} \tag{16.11}$$

The components of $\mathbf{e}^A$ can be expressed in terms of the displacement components by noting that

$$X_m = x_m - u_m \tag{16.12}$$

where now the displacements are regarded as functions of spatial position $\mathbf{x}$ (Eulerian description) rather than position in the reference configuration $\mathbf{X}$. Hence, the components of $\mathbf{F}^{-1}$ are $F^{-1}_{mn} = \partial X_m / \partial x_n$ or, in terms of the displacements,

$$F^{-1}_{mn} = \delta_{mn} - \frac{\partial u_m}{\partial x_n} \tag{16.13}$$

Substituting into (16.11) yields

$$e^A_{ij} = \frac{1}{2}\left\{\frac{\partial u_i}{\partial x_j} + \frac{\partial u_j}{\partial x_i} - \frac{\partial u_k}{\partial x_i}\frac{\partial u_k}{\partial x_j}\right\} \tag{16.14}$$

By comparison to the component expression for the Green–Lagrange strain (16.6), the sign of the last term is changed and the derivatives are with respect to position in the current configuration. Neglecting the last nonlinear terms reduces (16.14) to the expression for the small-strain tensor in which the distinction between the current and reference positions is neglected.

16.3   Relations Between D and Rates of $\mathbf{E}^G$ and U

16.3.1   Relation Between $\dot{\mathbf{E}}$ and D

Because $\mathbf{D}$ expresses the rate of deformation, we expect that there is a relation between $\mathbf{D}$ and the rate of strain, in particular the rate of the Green–Lagrange strain.

Differentiating the relation (15.2) yields an expression for $d\mathbf{v}$:

$$d\mathbf{v} = \dot{\mathbf{F}} \cdot d\mathbf{X}$$

Using (15.2) and comparing to (14.2) yields

$$\dot{\mathbf{F}} = \mathbf{L} \cdot \mathbf{F} \tag{16.15}$$

Differentiating the expression for the Green–Lagrange strain (16.3) gives

$$\dot{\mathbf{E}}^G = \frac{1}{2}\left\{\dot{\mathbf{F}}^T \cdot \mathbf{F} + \mathbf{F}^T \cdot \dot{\mathbf{F}}\right\}$$

Substituting (16.15) and rearranging gives

$$\dot{\mathbf{E}}^G = \mathbf{F}^T \cdot \mathbf{D} \cdot \mathbf{F} \tag{16.16}$$

Thus, $\dot{\mathbf{E}}^G = \mathbf{0}$ when $\mathbf{D} = \mathbf{0}$. This is a property of any material strain tensor (its rate vanishes when $\mathbf{D} = \mathbf{0}$) and, hence, reinforces the interpretation of a *material* strain measure.

Rates of the spatial strain measures do not vanish when $\mathbf{D}$ vanishes. For example, consider the rate of the Almansi strain (16.10)

$$\dot{\mathbf{e}}^A = -\frac{1}{2}\left\{ \frac{d}{dt}\left(\mathbf{F}^{-1T}\right) \cdot \mathbf{F}^{-1} + \mathbf{F}^{-1T} \cdot \frac{d}{dt}\left(\mathbf{F}^{-1}\right) \right\} \tag{16.17}$$

In order to calculate the rate of $\mathbf{F}^{-1}$, we begin with

$$\mathbf{F}^{-1} \cdot \mathbf{F} = \mathbf{I}$$

Differentiating and then solving for $d(\mathbf{F}^{-1})/dt$ gives

$$\frac{d}{dt}\left(\mathbf{F}^{-1}\right) = -\mathbf{F}^{-1} \cdot \dot{\mathbf{F}} \cdot \mathbf{F}^{-1}$$

This illustrates the general procedure for determining the derivative of the inverse of a tensor. Using (16.15) gives

$$\frac{d}{dt}\left(\mathbf{F}^{-1}\right) = -\mathbf{F}^{-1} \cdot \mathbf{L}$$

Substituting into (16.17) and rearranging gives

$$\dot{\mathbf{e}}^A = \mathbf{D} - \mathbf{L}^T \cdot \mathbf{e}^A - \mathbf{e}^A \cdot \mathbf{L}$$

When $\mathbf{D} = \mathbf{0}$, $\dot{\mathbf{e}}^A$ does not vanish but is

$$\dot{\mathbf{e}}^A = -\mathbf{W}^T \cdot \mathbf{e}^A - \mathbf{e}^A \cdot \mathbf{W}$$

Consequently, $\dot{\mathbf{e}}^A$ depends on the spin tensor and, in general, would not be suitable for use in a constitutive relation because the material behavior is affected by rigid rotation. This motivates the definition of a special rate that *does* vanish when $\mathbf{D} = \mathbf{0}$:

$$\hat{\mathbf{e}}^A = \dot{\mathbf{e}}^A + \mathbf{W}^T \cdot \mathbf{e}^A + \mathbf{e}^A \cdot \mathbf{W}$$

16.3.2   Relation Between $\mathbf{D}$ and $\dot{\mathbf{U}}$

We can also examine the relation between $\mathbf{D}$ and $\dot{\mathbf{U}}$. Rewriting (16.15), substituting (15.17), and rearranging yields

$$\mathbf{L} = \dot{\mathbf{R}} \cdot \mathbf{R}^T + \mathbf{R} \cdot \dot{\mathbf{U}} \cdot \mathbf{U}^{-1} \cdot \mathbf{R}^T \tag{16.18}$$

The first term $\dot{\mathbf{R}} \cdot \mathbf{R}^T$ is antisymmetric. Substituting (16.18) into (14.4) yields

$$\mathbf{D} = \frac{1}{2}\mathbf{R} \cdot \left\{ \dot{\mathbf{U}} \cdot \mathbf{U}^{-1} + \mathbf{U}^{-1T} \cdot \dot{\mathbf{U}}^T \right\} \cdot \mathbf{R}^T$$

Similarly, substituting into (14.5) yields

$$\mathbf{W} = \dot{\mathbf{R}} \cdot \mathbf{R}^T + \frac{1}{2}\mathbf{R} \cdot \left\{ \dot{\mathbf{U}} \cdot \mathbf{U}^{-1} - \mathbf{U}^{-1T} \cdot \dot{\mathbf{U}}^T \right\} \cdot \mathbf{R}^T$$

Although $\mathbf{U}$ is symmetric, in general the product $\dot{\mathbf{U}} \cdot \mathbf{U}^{-1}$ is not and, thus, the spin depends on the antisymmetric part of $\dot{\mathbf{U}} \cdot \mathbf{U}^{-1}$.

Exercises

16.1 Carry out the details leading from (16.11) to (16.14).

16.2 Carry out the details leading to (16.18).

16.3 Investigate the material strain measure $\mathbf{E}^{(-2)}$ corresponding to choice of the scale function $f(\Lambda) = \frac{1}{2}(1 - \Lambda^{-2})$.
   (a) Show that

$$\mathbf{E}^{(-2)} = \frac{1}{2}\left\{\mathbf{I} - \mathbf{C}^{-1}\right\}$$

   (b) Show that the Cartesian components of $\mathbf{E}^{(-2)}$ are given by

$$E_{ij}^{(-2)} = \frac{1}{2}\left(\frac{\partial u_i}{\partial x_j} + \frac{\partial u_j}{\partial x_i} - \frac{\partial u_i}{\partial x_k}\frac{\partial u_j}{\partial x_k} \right)$$

   and compare to the Cartesian component form of $\mathbf{e}^A$.
   (c) Show that $\mathbf{E}^{(-2)}$ is related to the Almansi strain measure $\mathbf{e}^A$ by

$$\mathbf{e}^A = \mathbf{R} \cdot \mathbf{E}^{(-2)} \cdot \mathbf{R}^T$$

16.4 Consider the spatial strain measure based on the scale function $g(\lambda) = \frac{1}{2}(\lambda^{-2} - 1)$.
   (a) Show that

$$\mathbf{e}^{(-2)} = \frac{1}{2}\left\{\mathbf{B} - \mathbf{I}\right\}$$

   (b) Show that the Cartesian form of $\mathbf{e}^{(-2)}$ is

$$e_{ij}^{(-2)} = \frac{1}{2}\left\{ \frac{\partial u_i}{\partial X_j} + \frac{\partial u_j}{\partial X_i} + \frac{\partial u_i}{\partial X_k}\frac{\partial u_j}{\partial X_k} \right\}$$

(c) Show that $e^{(-2)}$ is related to the Green–Lagrange strain measure E^G by

$$e^{(-2)} = R \cdot E^G \cdot R^T$$

16.5 By expanding $\ln \Lambda$ about $\Lambda = 1$ show that the logarithmic strain $E^{(\ln)}$ can be expressed as the following series:

$$E^{(\ln)} = (U - I) - \frac{1}{2}(U - I)^2 + \frac{1}{3}(U - I)^3 + \cdots$$

16.6 Show that the principal values of the Green–Lagrange strain measure E_K^G are related to those of the Almansi strain e_K^A by

$$e_K^A = \frac{E_K^G}{1 + 2E_K^G}$$

16.7 Beginning with $R \cdot R^T = I$, show that $\dot{R} \cdot R^T$ is antisymmetric.

16.8 Show that when the principal axes of U are fixed (do not rotate), D has the interpretation of the logarithmic strain rate in the current configuration.

16.9 Show that

$$\dot{J} = J \operatorname{tr} L = J \operatorname{tr} D$$

beginning with

$$J = \det(F) = \epsilon_{ijk} F_{i1} F_{j2} F_{k3}$$

[Hint: Write $\dot{J} = F_{p1}F_{q2}F_{r3}h_{pqr}$. Then show that h_{pqr} changes sign with the interchange of two indices, vanishes when two indices are equal, and that $h_{123} = \operatorname{tr} L$.]

16.10 Begin with Nanson's formula (15.12) and show that

$$\frac{d}{dt}(da) = \{\operatorname{tr} L - n \cdot L \cdot n\} \, da$$

16.11 (a) Show that

$$\dot{\Lambda}/\Lambda = n_i n_j D_{ij}$$

(b) Show that

$$\ddot{\Lambda}/\Lambda = n_i n_j Q_{ij} + \dot{n}_k \dot{n}_k$$

where Q_{ij} is the spatial gradient of the acceleration:

$$Q_{ij} = \frac{1}{2} \left\{ \frac{\partial a_i}{\partial x_j} + \frac{\partial a_j}{\partial x_i} \right\}$$

References

Biot MA 1965 *Mechanics of Incremental Deformation*. John Wiley & Sons, Ltd.
Hill R 1968 On constitutive inequalities for simple materials – I. *Journal of the Mechanics and Physics of Solids* **16**, 229–242.
Ogden RW 1997 *Non-Linear Elastic Deformations*. Dover.

17

Linearized Displacement Gradients

We now want to show that the deformation and large-strain measures reduce to the usual expressions for *small* strain when displacement gradients are infinitesimal. This is a useful exercise even though the result is expected because the large-strain measures have been constructed to have this property. The displacement is the difference between the positions in current and reference configurations $\mathbf{u} = \mathbf{x} - \mathbf{X}$ or, in component form, $u_k = x_k - X_k$.

The deformation gradient tensor (15.1) is then

$$F_{ij} = \frac{\partial x_i}{\partial X_j} = \delta_{ij} + \frac{\partial u_i}{\partial X_j} \tag{17.1}$$

or, in symbolic, coordinate-free form

$$\mathbf{F} = \mathbf{I} + (\nabla_{\mathbf{X}}\mathbf{u})^T \tag{17.2}$$

where $(\nabla_{\mathbf{X}}\mathbf{u})^T$ is the displacement gradient tensor and the subscript $\mathbf{X}$ emphasizes that the gradient is with respect to position in the reference configuration. We have shown that all the geometric measures of deformation, changes in the length of lines, changes in angles, and changes in volume can be expressed in terms of the Green deformation tensor $\mathbf{C}$ (15.6). Expressing $\mathbf{C}$ in terms of the displacement gradient yields

$$\mathbf{C} = \mathbf{I} + (\nabla_{\mathbf{X}}\mathbf{u})^T + (\nabla_{\mathbf{X}}\mathbf{u}) + (\nabla_{\mathbf{X}}\mathbf{u}) \cdot (\nabla_{\mathbf{X}}\mathbf{u})^T \tag{17.3}$$

or, in component form,

$$C_{ij} = \delta_{ij} + \left(\frac{\partial u_i}{\partial X_j} + \frac{\partial u_j}{\partial X_i} \right) + \frac{\partial u_k}{\partial X_i}\frac{\partial u_k}{\partial X_j} \tag{17.4}$$

We assume that the magnitude of the displacement gradient is much less than unity

$$\left| \frac{\partial u_i}{\partial X_j} \right| \ll 1 \tag{17.5}$$

Fundamentals of Continuum Mechanics, First Edition. John W. Rudnicki.
© 2015 John Wiley & Sons, Ltd. Published 2015 by John Wiley & Sons, Ltd.

and define the infinitesimal (small) strain tensor as

$$\varepsilon_{ij} = \frac{1}{2}\left(\frac{\partial u_i}{\partial X_j} + \frac{\partial u_j}{\partial X_i}\right) = \varepsilon_{ji} \tag{17.6}$$

or

$$\varepsilon = \frac{1}{2}\left(\nabla_{\mathbf{X}}\mathbf{u} + \left(\nabla_{\mathbf{X}}\mathbf{u}\right)^T\right) \tag{17.7}$$

Because of the assumption (17.5), the last terms in (17.3) and (17.4) can be neglected. Thus,

$$\mathbf{C} \approx \mathbf{I} + 2\varepsilon \tag{17.8}$$

or

$$C_{ij} \approx \delta_{ij} + 2\varepsilon_{ij} \tag{17.9}$$

Now we use these to linearize the geometric measures of deformation and express them in terms of the infinitesimal strain tensor.

17.1   Linearized Geometric Measures

17.1.1   Stretch in Direction $\mathbf{N}$

The stretch ratio in direction $\mathbf{N}$ is given by (15.7). Substituting (17.8) yields

$$\Lambda \approx \{\mathbf{N}\cdot(\mathbf{I}+2\varepsilon)\cdot\mathbf{N}\}^{1/2} = \sqrt{1+2\mathbf{N}\cdot\varepsilon\cdot\mathbf{N}}$$

Retaining only the linear term in the expansion

$$(1+x)^n = 1 + nx + \ldots \tag{17.10}$$

gives

$$\Lambda \approx 1 + \mathbf{N}\cdot\varepsilon\cdot\mathbf{N} \tag{17.11}$$

Therefore, to first order, the normal components of the infinitesimal strain tensor give the change in length of a line in the $\mathbf{N}$ direction in the reference configuration divided by its length in the reference configuration:

$$\mathbf{N}\cdot\varepsilon\cdot\mathbf{N} = \Lambda - 1$$

For example, if $\mathbf{N} = \mathbf{e}_1$, then ε_{11} is the change in length of a line segment originally in the X_1 direction divided by its original length.

17.1.2 Angle Change

The current angle between lines that were in directions $\mathbf{N}_A$ and $\mathbf{N}_B$ in the reference configuration is given by (15.9). Writing the current angle θ in terms of the angle in the reference configuration Θ and the change $\gamma = \Theta - \theta$ gives

$$\cos\{\Theta - \gamma\} = \frac{\mathbf{N}_A \cdot \mathbf{C} \cdot \mathbf{N}_B}{\Lambda_A \Lambda_B} \tag{17.12}$$

When $\mathbf{N}_A$ and $\mathbf{N}_B$ are orthogonal, i.e., $\mathbf{N}_A \cdot \mathbf{N}_B = 0$, (17.12) reduces to

$$\sin\gamma = \frac{\mathbf{N}_A \cdot \mathbf{C} \cdot \mathbf{N}_B}{\Lambda_A \Lambda_B} \tag{17.13}$$

where γ is the shear. Approximating $\sin\gamma$ by γ, substituting (17.8) and (17.11) into (17.13), and linearizing yields

$$\gamma \approx 2\mathbf{N}_A \cdot \boldsymbol{\varepsilon} \cdot \mathbf{N}_B \tag{17.14}$$

For example, if $\mathbf{N}_A = \mathbf{e}_1$ and $\mathbf{N}_B = \mathbf{e}_2$, $\gamma = 2\varepsilon_{12}$. Therefore, ε_{12} is one-half the change in angle between lines originally in the X_1 and X_2 directions.

17.1.3 Volume Change

The ratio of volume elements in the current and reference configurations is given by (15.14)

$$\frac{dv}{dV} = \det(\mathbf{F})$$

Substituting (17.1) and expanding the determinant yields

$$\frac{dv}{dV} = \epsilon_{ijk}\left(\delta_{i1} + \frac{\partial u_i}{\partial X_1}\right)\left(\delta_{j2} + \frac{\partial u_j}{\partial X_2}\right)\left(\delta_{k3} + \frac{\partial u_k}{\partial X_3}\right)$$

Carrying out the multiplication but keeping only the linear terms in the displacement gradient components gives

$$\frac{dv}{dV} \approx 1 + \frac{\partial u_1}{\partial X_1} + \frac{\partial u_2}{\partial X_2} + \frac{\partial u_3}{\partial X_3} = 1 + \varepsilon_{11} + \varepsilon_{22} + \varepsilon_{33}$$

Thus, the change in volume divided by reference volume is the trace of the small-strain tensor.

17.2   Linearized Polar Decomposition

The polar decomposition is given by (15.17), $\mathbf{F} = \mathbf{R} \cdot \mathbf{U}$, where $\mathbf{F}$ is given by (17.1) or (17.2). To determine the linearized form of $\mathbf{U} = \sqrt{\mathbf{C}}$, we begin by expressing the approximation of $\mathbf{C}$, (17.8) or (17.9), in principal axis form

$$\mathbf{C} \approx (1 + 2\varepsilon_I)\mathbf{N}_I\mathbf{N}_I + (1 + 2\varepsilon_{II})\mathbf{N}_{II}\mathbf{N}_{II} + (1 + 2\varepsilon_{III})\mathbf{N}_{III}\mathbf{N}_{III} \tag{17.15}$$

and obtaining

$$\mathbf{U} = \sqrt{\mathbf{C}} \approx \sqrt{1 + 2\varepsilon_I}\,\mathbf{N}_I\mathbf{N}_I + \ldots \tag{17.16}$$

Linearizing using (17.10) then yields

$$\mathbf{U} = (1 + \varepsilon_I)\mathbf{N}_I\mathbf{N}_I + \ldots \tag{17.17}$$

Reverting to coordinate-free form yields

$$\mathbf{U} \approx \mathbf{I} + \boldsymbol{\varepsilon} \tag{17.18}$$

or, in terms of Cartesian components,

$$U_{ij} \approx \delta_{ij} + \varepsilon_{ij} \tag{17.19}$$

The linearized form of the rotation tensor is determined from

$$\mathbf{R} = \mathbf{F} \cdot \mathbf{U}^{-1} \tag{17.20}$$

By means of the same procedure as in (17.15) to (17.19), the linearized form of $\mathbf{U}^{-1}$ is

$$\mathbf{U}^{-1} \approx \mathbf{I} - \boldsymbol{\varepsilon} \tag{17.21}$$

Substituting (17.21) and (17.2) into (17.20) and neglecting second-order terms gives

$$\mathbf{R} \approx (\mathbf{I} + (\boldsymbol{\nabla}_{\mathbf{X}}\mathbf{u})^T) \cdot (\mathbf{I} - \boldsymbol{\varepsilon})$$

$$\approx \mathbf{I} + \frac{1}{2}((\boldsymbol{\nabla}_{\mathbf{X}}\mathbf{u})^T - (\boldsymbol{\nabla}_{\mathbf{X}}\mathbf{u}))$$

The final term is the *infinitesimal rotation tensor*

$$\boldsymbol{\Omega} = \frac{1}{2}\left[(\boldsymbol{\nabla}_{\mathbf{X}}\mathbf{u})^T - \boldsymbol{\nabla}_{\mathbf{X}}\mathbf{u}\right] \tag{17.22}$$

or in component form

$$\Omega_{ij} = \frac{1}{2}\left(\frac{\partial u_i}{\partial X_j} - \frac{\partial u_j}{\partial X_i}\right) \tag{17.23}$$

Thus, the multiplicative decomposition (15.17) reduces to the additive decomposition of the displacement gradient tensor into the symmetric infinitesimal strain tensor and the skew-symmetric infinitesimal rotation tensor

$$(\nabla_X \mathbf{u})^T = \varepsilon + \Omega \tag{17.24}$$

17.3   Small-Strain Compatibility

If the displacements $u_k(X_j, t)$ are known and differentiable, then it is always possible to compute the six strain components

$$\varepsilon_{ij} = \frac{1}{2} \left(\frac{\partial u_i}{\partial X_j} + \frac{\partial u_j}{\partial X_i} \right)$$

Because there are six strain components calculated from three displacements, some relations must exist between the strain components. That is, the small-strain components must be *compatible*. A mathematically analogous, but simpler situation occurs when force components P_i are calculated from a scalar potential ϕ:

$$\mathbf{P} = \nabla\phi \tag{17.25}$$

where the gradient is with respect to displacement components u_i. In general, the force components are independent, but if they satisfy (17.25) then they must also satisfy

$$\nabla \times \mathbf{P} = 0 \tag{17.26}$$

This requires, for example, that

$$\frac{\partial P_1}{\partial u_2} = \frac{\partial P_2}{\partial u_1}$$

This condition is obtained from the X_3 component of (17.26) or by substituting the force components from (17.25) and noting that the derivatives of ϕ with respect to X_1 and X_2 may be taken in either order.

The equations of small-strain compatibility can be obtained in similar fashion by differentiating the strain components, writing them in terms of displacements, and interchanging the order of differentiation. An example is the following:

$$2\varepsilon_{12,12} = u_{1,212} + u_{2,112}$$

$$= u_{1,122} + u_{2,211}$$

$$= \varepsilon_{11,22} + \varepsilon_{22,11}$$

where, for brevity, derivatives are denoted by $\partial u_i / \partial X_j = u_{i,j}$ etc. Similar manipulations yield five additional conditions:

$$\varepsilon_{22,33} + \varepsilon_{33,22} - 2\varepsilon_{23,23} = 0$$
$$\varepsilon_{33,11} + \varepsilon_{11,33} - 2\varepsilon_{31,31} = 0$$
$$-\varepsilon_{11,23} + \left(-\varepsilon_{23,1} + \varepsilon_{31,2} + \varepsilon_{12,3}\right)_{,1} = 0$$
$$-\varepsilon_{22,13} + \left(\varepsilon_{23,1} - \varepsilon_{31,2} + \varepsilon_{12,3}\right)_{,2} = 0$$
$$-\varepsilon_{33,12} + \left(\varepsilon_{23,1} + \varepsilon_{31,2} - \varepsilon_{12,3}\right)_{,3} = 0$$

All six can be summarized concisely as

$$\mathbf{\nabla_X} \times (\mathbf{\nabla_X} \times \boldsymbol{\varepsilon})^T = 0 \tag{17.27}$$

or

$$\epsilon_{jrs}\epsilon_{ipq}\varepsilon_{sq\cdot rp} = 0 \tag{17.28}$$

Malvern (1988, p. 187) explains that of these six only three are independent. Nevertheless, it is generally more convenient to use all six but recognize that they provide only three independent conditions.

Thus, if the strains are written in terms of displacements, the conditions (17.27) or (17.28) are *necessary* for the strains to be compatible. On the other hand, if the strains are known, what conditions are *sufficient* to guarantee that the strain components can be integrated to yield a single-valued displacement field? To visualize the meaning of this, imagine cutting the body into small (infinitesimal) blocks. Assign a strain to each block. Generally the body will not fit back together. There will be gaps, overlaps, etc. That is, the displacement field will not be single valued unless the strains assigned to the blocks are *compatible*. It turns out that the conditions (17.28) are also sufficient (at least in simply connected bodies; if the body is not simply connected additional conditions are needed).

Again the situation is mathematically analogous to a simpler one. Consider the increment of work dW due to the action of the force $\mathbf{P}$ on the displacement increment $d\mathbf{u}$

$$dW = \mathbf{P} \cdot d\mathbf{u}$$

In general, dW is not a perfect differential. That is, work is a path-dependent quantity and the line integral

$$W = \int_C \mathbf{P} \cdot d\mathbf{u} \tag{17.29}$$

will have different values if calculated on different paths between the same two points. It follows that the integral around a closed path will not be zero. Work will, however, be path independent if it is equal to the change in energy, or if, in other words, the system is conservative. A condition guaranteeing that this is the case is the same as (17.26)

$$\mathbf{\nabla} \times \mathbf{P} = 0 \tag{17.30}$$

If this condition is met, the force can be represented as the gradient of a scalar potential function (17.25). Hence, (17.30) is *necessary* and *sufficient* for the force to be the gradient of a scalar function and the work to be equal to a change in energy.

The situation is similar but the details are more complicated for strain compatibility because the strain is a tensor. Consider the conditions for which the displacement gradient field can be integrated to give a single-valued displacement field

$$\mathbf{u}^P - \mathbf{u}^O = \int_C d\mathbf{u} = \int_C (\nabla_\mathbf{X} \mathbf{u})^T \cdot d\mathbf{X} \tag{17.31}$$

where $\mathbf{u}^P$ is the displacement at point P, $\mathbf{u}^O$ is the displacement at O, and C is any path joining P and O. Using (17.24) and expressing in index notation, (17.31) becomes

$$u_i^P - u_i^O = \int_C (\epsilon_{ij} + \Omega_{ij}) dX_j$$

Expressing the second term on the right in terms of the infinitesimal rotation vector $\Omega_{ij} = \epsilon_{jik} w_k$ yields

$$u_i^P - u_i^O = \int_C (\epsilon_{ij} + \epsilon_{jik} w_k) dX_j \tag{17.32}$$

Analogous to (17.29) and (17.30), a sufficient condition guaranteeing that the integral (17.32) is independent of path is that the curl of the integrand must vanish. This operation yields

$$\epsilon_{ipq} \epsilon_{qs,p} + w_{i,s} - \delta_{is} w_{p,p} = 0 \tag{17.33}$$

but the second term on the right side vanishes because the divergence of the rotation vector is zero. Operating on both sides with $\epsilon_{jrs} \partial_r$ yields (17.28).

Exercises

17.1 Fill in the details of linearizing (17.13) to get (17.14).

17.2 Fill in the details of obtaining (17.21).

17.3 Derive the linearized form of Nanson's formula (15.12) relating area elements in the current and reference configurations.

17.4 Show that the divergence of the rotation vector vanishes.

17.5 Show that applying the curl to the integrand of (17.32) and setting equal to zero results in (17.33).

Reference

Malvern LE 1988 *Introduction to the Mechanics of a Continuous Medium*. Prentice Hall.

Part Four

Balance of Mass, Momentum, and Energy

In this part we develop forms for the conservation of mass, momentum, and energy appropriate for application to a continuum. The starting point is application of these laws to a group of materials or element of mass that involves integration over volume elements. Consequently, preparatory to derivation of the balance laws, we need to discuss manipulation of the types of integrals that occur.

Fundamentals of Continuum Mechanics, First Edition. John W. Rudnicki.
© 2015 John Wiley & Sons, Ltd. Published 2015 by John Wiley & Sons, Ltd.

18

Transformation of Integrals

To derive equations expressing the conservation of mass and energy and balance of angular and linear momentum, we will repeatedly use the divergence theorem or Green–Gauss theorem. The theorem relates the integral of the divergence over a volume to an integral over the bounding surface with outward normal $\mathbf{n}$. For a vector $\mathbf{u}$ the theorem is

$$\int_V \nabla \cdot \mathbf{u} \, dV = \int_S \mathbf{n} \cdot \mathbf{u} \, dS \tag{18.1}$$

where V is the volume, S is the bounding surface, and $\mathbf{n}$ is the unit outward normal on S. The following related theorems for a scalar function ϕ and a tensor function $\mathbf{F}$ have the same form:

$$\int_V \nabla \phi \, dV = \int_S \mathbf{n}\phi \, dS$$

$$\int_V \nabla \cdot \mathbf{F} \, dV = \int_S \mathbf{n} \cdot \mathbf{F} \, dS \tag{18.2}$$

Aris (1989) and Kellogg (1954), or other books on potential theory have extensive discussions of this theorem. Here, for simplicity, we will consider a planar version:

$$\int\int_A \left(\frac{\partial u_x}{\partial x} + \frac{\partial u_y}{\partial y} \right) dx \, dy = \int_C (n_x u_x + n_y u_y) \, ds \tag{18.3}$$

As shown in Figure 18.1, the curve C, composed of segments C_1 and C_2, encloses the area A and is traversed in a counterclockwise direction. Figure 18.2 shows the components of the outward normal and an element ds of the curve C. Although it is usually written in different form, this is Green's theorem in the plane.

To prove this theorem note that the double integral of the second term on the left of (18.3) can be carried out by first integrating in y for a vertical strip of width dx (Figure 18.1). The limits of integration are given by the curves $y_1(x)$ and $y_2(x)$ that make up C_1 and C_2. Then the integration in x is carried out by sweeping this strip from left to right. Because $\partial u_y / \partial y$ is a

Fundamentals of Continuum Mechanics, First Edition. John W. Rudnicki.
© 2015 John Wiley & Sons, Ltd. Published 2015 by John Wiley & Sons, Ltd.

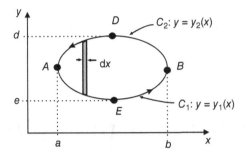

Figure 18.1 Definitions for derivation of Green's theorem in the plane.

perfect differential, the integration in y is simply

$$\int\int_A \frac{\partial u_y}{\partial y} dx \, dy = \int_a^b dx \int_{y_1(x)}^{y_2(x)} \frac{\partial u_y}{\partial y}(x,y) \, dy$$

$$= \int_a^b [u_y(x,y_2(x)) - u_y(x,y_1(x))] \, dx$$

$$= -\int_b^a u_y(x,y_2(x)) \, dx - \int_a^b u_y(x,y_1(x)) \, dx$$

$$= -\int_C u_y \, dx$$

The third line follows by inserting a minus sign and interchanging the limits of integration in the first term. The last line follows by noting that the sum of integrating over the curves $y_1(x)$ and $y_2(x)$ in the same direction is an integral around the closed curve C. Writing

$$dx = \frac{dx}{ds} ds = -\sin \alpha \, ds$$

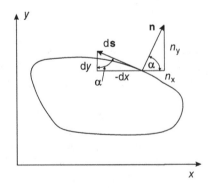

Figure 18.2 Expressing Green's theorem in the plane in terms of the normal and tangent vectors to the curve.

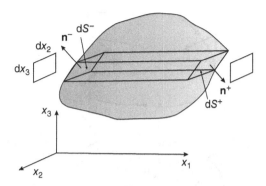

Figure 18.3   Schematic of integration over a columnar element in the x_1 direction.

and noting that $n_y = \sin \alpha$ gives the second term on the right side of (18.3). Integration of the first term follows in the same way but by first using a horizontal strip of height dy.

Proofs of the three-dimensional version (18.1) are similar but use integration along columnar elements as illustrated in Figure 18.3. Integration in the x_1 direction becomes

$$\int_V \frac{\partial}{\partial x_1} (\ldots) \, dx_1 \, dx_2 \, dx_3 = \int_{S+} (\ldots)^+ \, dx_2 \, dx_3 - \int_{S-} (\ldots)^- \, dx_2 \, dx_3$$

where the right side is the contribution from the ends of the column. Noting that

$$dx_2 \, dx_3 = n_1^+ \, ds^+ = -n_1^- \, ds^-$$

and adding the contributions from all the columns gives

$$\int_V \frac{\partial}{\partial x_1} (\ldots) \, dx_1 \, dx_2 \, dx_3 = \int_S n_1 \, (\ldots) \, ds$$

Treating the partial derivatives with respect to x_2 and x_3 in the same way and adding the results establishes the theorem.

The curve in Figure 18.1 is a special one because vertical and horizontal lines intersect the curve in no more than two points. Nevertheless the theorem applies for more complicated curves such as those shown in Figure 18.4 and the method of proof used above is easily modified for these cases. For the curve on the left a vertical line can intersect the curve in four points. This difficulty is easily overcome, however, by inserting the dotted line as shown and applying the method to each part of the area separately. The dotted line is traversed in opposite directions for each part and, thus, as long as the integrand is continuous, the contributions cancel.

On the right in Figure 18.4, the area of integration A has a hole so that there is an interior and exterior boundary. Again, demonstration of the theorem proceeds in the same way after connecting the interior and exterior boundaries by the dotted line. If the integrand is continuous, the portions of the integral over the dotted line cancel since they are traversed in opposite directions. Note that the resulting contour C is counterclockwise on the exterior boundary and

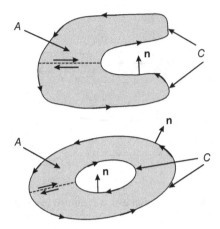

Figure 18.4 Curves for which vertical or horizontal lines intersect the boundaries in more than two points.

clockwise on the interior boundary. On both boundaries the normal **n** points out of the area A. In other words, a person walking on the contour in the direction shown would have the area A to their left and the normal **n** to their right.

A similar argument can be applied to integration over columns illustrated in Figure 18.3.

Exercises

18.1 Carry out integration of the first term on the left side in (18.3) to obtain the first term on the right side.

18.2 Express the surface integral over a closed surface S as a volume integral over the enclosed volume V if the integrand of the surface integral is:
   (a) $\mathbf{n} \cdot \boldsymbol{\sigma} \cdot \mathbf{v}$
   (b) $\epsilon_{rms} x_m \sigma_{js} n_j$

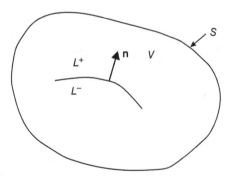

Figure 18.5 Volume V with a bounding surface S. L is a surface of discontinuity with normal **n** and positive and negative sides as shown.

18.3 Modify the divergence theorem applied to the stress tensor σ

$$\int_V \nabla \cdot \sigma \, dV = \int_S \mathbf{n} \cdot \sigma \, dS$$

for the case in which the traction $\mathbf{t} = \mathbf{n} \cdot \sigma$ is discontinuous across an internal surface L contained in V as shown in Figure 18.5. (Be sure to define all terms that enter.)

References

Aris R 1989 *Vectors, Tensors, and the Basic Equations of Fluid Mechanics*. Dover.
Kellogg OD 1954 *Foundations of Potential Theory*. Dover.

19

Conservation of Mass

In this chapter, we derive equations expressing the conservation of mass for a continuum. The methods we develop here will be used in succeeding chapters to derive equations for the balance of momentum and conservation of energy.

The total mass in a reference volume V is

$$m = \int_V \rho_o(\mathbf{X}) \, dV$$

In the current configuration, this same mass occupies the volume v:

$$m = \int_V \rho_o(\mathbf{X}) \, dV = \int_{v(t)} \rho(\mathbf{x}, t) dv \tag{19.1}$$

Because mass can be neither created nor destroyed the rate of change of mass must vanish: $dm/dt = 0$. Differentiating (19.1) yields

$$\frac{d}{dt} \int_{v(t)} \rho(\mathbf{x}, t) \, dv = 0 \tag{19.2}$$

Because the current volume v occupied by a fixed amount of mass changes with time, the integration volume in (19.2) depends on time. In Section 19.2 we discuss how to compute directly the derivative of an integral over a time-dependent volume. First, however, we introduce another approach by converting the integral to one over the reference volume. Since the current and reference volume elements are related by $dv = J \, dV$ where $J = \det(\mathbf{F})$, we can rewrite (19.2) as an integral over the reference volume

$$\frac{d}{dt} \int_V \rho \, [\mathbf{x}(\mathbf{X}, t), t] J \, dV = 0$$

Fundamentals of Continuum Mechanics, First Edition. John W. Rudnicki.
© 2015 John Wiley & Sons, Ltd. Published 2015 by John Wiley & Sons, Ltd.

The integration variable is now position in the reference configuration $\mathbf{X}$ rather than position in the current configuration $\mathbf{x}$, and J is the Jacobian of the change of variable. Because the reference volume is independent of time, we can take the derivative inside the integral

$$\int_V \left\{ J\frac{d}{dt}\rho + \rho\dot{J} \right\} dV = 0 \tag{19.3}$$

Problem 16.9 outlines how to compute the derivative of the Jacobian J. The result is

$$\dot{J} = J\mathrm{tr}\mathbf{D} = J\mathrm{tr}\mathbf{L} = J\nabla \cdot \mathbf{v} \tag{19.4}$$

Substituting into (19.3) yields

$$\int_V \left\{ \frac{d}{dt}\rho + \rho\nabla \cdot \mathbf{v} \right\} J\, dV = 0$$

Now the integration can be changed back to the current volume

$$\int_{v(t)} \left\{ \frac{d\rho}{dt} + \rho\nabla \cdot \mathbf{v} \right\} dv = 0$$

Using the expression for the material derivative of the density, (13.12) or (13.11), we rewrite the integrand as

$$\int_{v(t)} \left\{ \frac{\partial\rho}{\partial t} + \nabla \cdot (\rho\mathbf{v}) \right\} dv = 0 \tag{19.5}$$

Using the divergence theorem on the second term gives

$$\int_{v(t)} \frac{\partial\rho}{\partial t}\, dv + \int_{a(t)} \mathbf{n} \cdot \mathbf{v}\, \rho\, da = 0 \tag{19.6}$$

Because $\partial/\partial t$ is the time derivative with the spatial position fixed, it can be taken outside the integral. Thus, the first term is the rate of change of mass instantaneously inside the spatial volume $v(t)$.

The second term in (19.6) is the rate of change of mass in v due to flow across the surface of v, i.e., a (Figure 19.1). Since $\mathbf{n}$ is the outward normal, the integral is positive for flow outward. During a time increment Δt the mass passing through da sweeps out a cylindrical volume

$$dv = \mathbf{n} \cdot \mathbf{v}\Delta t\, da \tag{19.7}$$

where $\mathbf{v}$ is the material velocity. Therefore the mass outflow is $\rho\mathbf{v} \cdot \mathbf{n}\Delta t$ da. If we had begun the derivation by considering a control volume fixed in space, then the result would had the same form as (19.6). Although the concepts of following a fixed amount of mass or considering the change of mass in a control volume are different, the results must be the same.

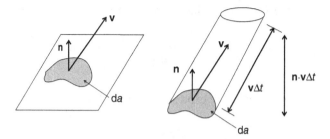

Figure 19.1   Illustration of the flux across a surface element da.

Because (19.5) applies for all volumes v containing a fixed amount of mass, the integrand vanishes and

$$\frac{\partial \rho}{\partial t} + \nabla \cdot (\rho \mathbf{v}) = 0 \qquad (19.8)$$

is the local form of mass conservation in the current configuration. This equation can be written in several alternative forms. Expanding the second term gives

$$\frac{\partial \rho}{\partial t} + \mathbf{v} \cdot \nabla \rho + \rho \nabla \cdot \mathbf{v} = 0$$

Using the material derivative gives

$$\frac{d\rho}{dt} + \rho \nabla \cdot \mathbf{v} = 0 \qquad (19.9)$$

Rearranging gives

$$\frac{1}{\rho}\frac{d\rho}{dt} = -\nabla \cdot \mathbf{v} \qquad (19.10)$$

The left side of (19.10) is the fractional rate of volume decrease. The right side gives an interpretation of the divergence as the flux out of a volume. Using (19.4) we can rewrite (19.10) as

$$\frac{d}{dt}(\rho J) = 0 \qquad (19.11)$$

Integrating yields

$$\rho J = \text{constant} = \rho_0$$

This is a local expression of (19.1).

For an incompressible material $d\rho/dt = 0$, not $\partial \rho/\partial t = 0$. (Note that to say a material is incompressible does not mean that it is rigid (non-deformable). The material can deform but in such a way that the volume remains constant.) It follows that for an incompressible material

$$\mathbf{\nabla} \cdot \mathbf{v} = 0 \tag{19.12}$$

In this case the velocity vector $\mathbf{v}$ can be expressed as the curl of a vector $\mathbf{\Psi}$:

$$\mathbf{v} = \mathbf{\nabla} \times \mathbf{\Psi}$$

A velocity of this form automatically satisfies (19.12).

19.1   Reynolds' Transport Theorem

In examining the other balance laws, we will encounter the derivative of integrals of the following form over a volume in the current configuration containing a fixed amount of mass

$$I = \frac{d}{dt} \int_{v(t)} \rho(\mathbf{x}, t) \, \mathcal{A}(\mathbf{x}, t) \, dv \tag{19.13}$$

where $\mathcal{A}(\mathbf{x}, t)$ is a vector or scalar quantity that is proportional to the mass, e.g., kinetic energy per unit mass, momentum per unit mass. As in the preceding section, the complications of differentiating an integral over a time-dependent volume are circumvented by converting to integration over the reference volume

$$I = \frac{d}{dt} \int_{V} \rho(\mathbf{x}[\mathbf{X}, t], t) \, \mathcal{A}(\mathbf{x}[\mathbf{X}, t], t) \, J \, dV \tag{19.14}$$

Now, the derivative can be taken inside the integral

$$I = \int_{V} \left\{ J \rho(\mathbf{x}[\mathbf{X}, t], t) \frac{d}{dt} \mathcal{A}(\mathbf{x}[\mathbf{X}, t], t) \right\} dV \tag{19.15}$$

where the other terms vanish because of mass conservation in the form (19.11). Converting the integral back to the current volume gives

$$I = \int_{v(t)} \rho(\mathbf{x}, t) \frac{d}{dt} \mathcal{A}(\mathbf{x}, t) \, dv \tag{19.16}$$

Equating (19.13) and (19.16) yields

$$\frac{d}{dt} \int_{v(t)} \rho(\mathbf{x}, t) \mathcal{A}(\mathbf{x}, t) \, dv = \int_{v(t)} \rho(\mathbf{x}, t) \frac{d}{dt} \mathcal{A}(\mathbf{x}, t) \, dv$$

Hence, if the integral is over a volume that encloses a fixed amount of mass, the material derivative can be taken inside the integral to operate only on $\mathcal{A}(\mathbf{x}, t)$. This is *Reynolds' transport theorem.*

19.2 Derivative of an Integral over a Time-Dependent Region

An alternative approach to dealing with integrals in (19.2) that are over a time-dependent volume is to fall back on the fundamental definition of a derivative. Because the volume of integration encloses a fixed set of material particles, the volume changes with time. We must take this into account in computing the derivative. Let $v(t)$ be the time-dependent volume and $\mathbf{n} \cdot \mathbf{v}$ be the normal speed of points on the boundary of v. We want to compute

$$\frac{d}{dt} \int_{v(t)} Q(\mathbf{x}, t) \, dv = \lim_{\Delta t \to 0} \frac{1}{\Delta t} \left\{ \int_{v(t+\Delta t)} Q(\mathbf{x}, t + \Delta t) \, dv - \int_{v(t)} Q(\mathbf{x}, t) \, dv \right\} \quad (19.17)$$

where $Q(\mathbf{x}, t)$ is the spatial description of some quantity defined everywhere in $v(t)$. We can write the volume at $t + \Delta t$ as

$$v(t + \Delta t) = v(t) + [v(t + \Delta t) - v(t)]$$

Therefore, (19.17) becomes

$$\frac{d}{dt} \int_{v(t)} Q(\mathbf{x}, t) \, dv = \lim_{\Delta t \to 0} \left\{ \frac{1}{\Delta t} \int_{v(t)} \{Q(\mathbf{x}, t + \Delta t) - Q(\mathbf{x}, t)\} \, dv \right\}$$

$$+ \lim_{\Delta t \to 0} \left\{ \frac{1}{\Delta t} \int_{v(t+\Delta t)-v(t)} Q(\mathbf{x}, t + \Delta t) \, dv \right\}$$

Because the first integral does not depend on Δt, the limit can be taken inside the integral to yield

$$\int_{v(t)} \lim_{\Delta t \to 0} \left\{ \frac{Q(\mathbf{x}, t + \Delta t) - Q(\mathbf{x}, t)}{\Delta t} \right\} \, dv$$

Using the definition of the partial derivative then gives

$$\int_{v(t)} \frac{\partial Q}{\partial t} (\mathbf{x}, t) \, dv$$

To evaluate the second term, consider the motion of a portion of the boundary (Figure 19.1). Equation (19.7) is the volume swept out in time Δt. Therefore

$$\frac{d}{dt} \int_{v(t)} Q(\mathbf{x}, t) \, dv = \int_{v(t)} \frac{\partial Q}{\partial t} (\mathbf{x}, t) \, dv + \int_{a(t)} Q(\mathbf{x}, t) \mathbf{n} \cdot \mathbf{v} \, da$$

The last term can be transformed using the divergence theorem applied to a control volume instantaneously coinciding with the volume occupied by the material. Thus, the final result is

$$\frac{d}{dt} \int_{v(t)} Q(\mathbf{x}, t) \, dv = \int_{v(t)} \left\{ \frac{\partial Q}{\partial t}(\mathbf{x}, t) + \nabla \cdot [Q(\mathbf{x}, t)\mathbf{v}] \right\} dv \qquad (19.18)$$

If $Q(\mathbf{x}, t) = \rho(\mathbf{x}, t)$ is the density, the left side of (19.18) vanishes because of mass conservation. Hence, the right-hand side must also vanish and, since the equation must apply for any volume v, (19.8) is recovered.

If $Q(\mathbf{x}, t) = 1$, (19.18) becomes

$$\frac{d}{dt} \int_{v(t)} dv = \int_{v(t)} \nabla \cdot \mathbf{v} \, dv = \int_{a(t)} \mathbf{n} \cdot \mathbf{v} \, da \qquad (19.19)$$

reinforcing the interpretation of the divergence as a measure of the flux out of a volume.

19.3   Example: Mass Conservation for a Mixture

Mixture theory assumes that each material point can be simultaneously occupied by more than one constituent. Consider a mixture of two phases, a solid with density ρ_s and a fluid with density ρ_f. If there are no chemical reactions between the solid and fluid phases then separate mass conservation equations for each species take the form

$$\frac{\partial \rho_s}{\partial t} + \nabla \cdot (\rho_s \mathbf{v}_s) = 0 \qquad (19.20)$$

$$\frac{\partial \rho_f}{\partial t} + \nabla \cdot (\rho_f \mathbf{v}_f) = 0 \qquad (19.21)$$

where $\mathbf{v}_s$ is the velocity of the solid and $\mathbf{v}_f$ is the velocity of the fluid. If there are chemical reactions converting the solid to fluid phase and vice versa, each equation will have a term of equal magnitude and opposite sign.

If the solid has a structure or framework, it is more convenient to combine these equations into a single one expressing mass conservation for the fluid in terms of flow *relative to the solid phase*. To this end we rewrite (19.20) as

$$\frac{d}{dt}\rho_s + \rho_s \nabla \cdot \mathbf{v}_s = 0 \qquad (19.22)$$

where

$$\frac{d}{dt}(\ldots) = \frac{\partial}{\partial t}(\ldots) + \mathbf{v}_s \cdot \nabla(\ldots)$$

is the material derivative following the motion of the solid. The second equation can be rewritten as

$$\frac{\partial}{\partial t}\rho_f + \mathbf{v}_s \cdot \nabla\rho_f + (\mathbf{v}_f - \mathbf{v}_s) \cdot \nabla\rho_f + \rho_f\nabla \cdot \mathbf{v}_f = 0$$

where $d(\ldots)/dt$ is again the material derivative following the motion of the solid. Adding and subtracting $\rho_f\nabla \cdot (\mathbf{v}_f - \mathbf{v}_s)$ gives

$$\frac{d}{dt}\rho_f + \nabla \cdot [\rho_f(\mathbf{v}_f - \mathbf{v}_s)] - \rho_f\nabla \cdot (\mathbf{v}_f - \mathbf{v}_s) + \rho_f\nabla \cdot \mathbf{v}_f = 0$$

Cancelling the term $\rho_f\nabla \cdot \mathbf{v}_f$ leaves

$$\frac{d}{dt}\rho_f + \nabla \cdot [\rho_f(\mathbf{v}_f - \mathbf{v}_s)] + \rho_f\nabla \cdot \mathbf{v}_s = 0$$

Using (19.22) to replace $\nabla \cdot \mathbf{v}_s$ yields

$$\frac{\partial m}{\partial t} + \nabla \cdot \mathbf{q} = 0$$

where

$$\mathbf{q} = \rho_f(\mathbf{v}_f - \mathbf{v}_s)$$

is the flux of fluid mass relative to the solid and

$$\frac{\partial m}{\partial t} = \rho_f\left[\frac{1}{\rho_f}\frac{d\rho_f}{dt} - \frac{1}{\rho_s}\frac{d\rho_s}{dt}\right]$$

is the change of fluid mass relative to the change of solid volume.

Exercises

19.1 Fill in the details of obtaining (19.15) from (19.14).

19.2 Evaluate (19.19) for a spherical volume with time-dependent radius $R(t)$.

20

Conservation of Momentum

20.1   Momentum Balance in the Current State

20.1.1   Linear Momentum

The conservation of linear momentum expresses the application of Newton's second law:

$$\sum \mathbf{F} = \frac{\mathrm{d}}{\mathrm{d}t}(m\mathbf{v}) \tag{20.1}$$

To apply this to a continuum, we enforce the condition that the momentum be balanced for a set of particles with a fixed mass. An alternative approach, as indicated in the preceding chapter for the conservation of mass, is to focus on a control volume fixed in space but to account for the flux of momentum in and out of the volume. (See Problem 20.3.) Let $\mathbf{t}$ be the surface force per unit current area, that is, the traction. Let $\mathbf{b}$ be the body force per unit mass. Application of (20.1) to a volume $v(t)$ enclosed by a surface $a(t)$ gives

$$\int_{a(t)} \mathbf{t}\,\mathrm{d}a + \int_{v(t)} \rho\mathbf{b}\,\mathrm{d}v = \frac{\mathrm{d}}{\mathrm{d}t}\int_{v(t)} \rho\mathbf{v}\,\mathrm{d}v \tag{20.2}$$

Writing the traction in terms of the stress as $\mathbf{n}\cdot\boldsymbol{\sigma}$ and using the divergence theorem on the first term yields

$$\int_{a(t)} \mathbf{n}\cdot\boldsymbol{\sigma}\,\mathrm{d}a = \int_{v(t)} \boldsymbol{\nabla}\cdot\boldsymbol{\sigma}\,\mathrm{d}v \tag{20.3}$$

Alternatively, conservation of linear momentum can be used to define the stress as that tensor necessary to convert the surface integral in (20.2) into a volume integral. Reynolds' transport theorem gives the following result for the right hand side of (20.2):

$$\int_{v(t)} \rho\frac{\mathrm{d}\mathbf{v}}{\mathrm{d}t}\,\mathrm{d}v$$

Fundamentals of Continuum Mechanics, First Edition. John W. Rudnicki.
© 2015 John Wiley & Sons, Ltd. Published 2015 by John Wiley & Sons, Ltd.

Collecting terms gives

$$\int_{v(t)} \left\{ \boldsymbol{\nabla} \cdot \boldsymbol{\sigma} + \rho\mathbf{b} - \rho\frac{d\mathbf{v}}{dt} \right\} dv = 0$$

Since this integral must vanish for *any* material volume, the integrand must vanish:

$$\boldsymbol{\nabla} \cdot \boldsymbol{\sigma} + \rho\mathbf{b} = \rho\frac{d\mathbf{v}}{dt}$$

or, in component form,

$$\frac{\partial \sigma_{ij}}{\partial x_i} + \rho b_j = \rho\frac{dv_j}{dt} \tag{20.4}$$

This is the *equation of motion.* If the right hand side is negligible, then (20.4) reduces to the *equilibrium equation*

$$\frac{\partial \sigma_{ij}}{\partial x_i} + \rho b_j = 0$$

expressing that the sum of the forces is zero.

20.1.2 Angular Momentum

Similarly, balance of angular momentum results from the statement that the sum of the moments **M** is equal to the time derivative of the angular momentum **L**

$$\sum \mathbf{M} = \frac{d}{dt}\mathbf{L}$$

Applying this to a collection of material particles occupying the current volume v(t) enclosed by the surface a(t) yields

$$\int_{a(t)} (\mathbf{x} \times \mathbf{t})\, da + \int_{v(t)} (\mathbf{x} \times \rho\mathbf{b})\, dv = \frac{d}{dt}\int_{v(t)} (\mathbf{x} \times \rho\mathbf{v})\, dv$$

or in component form

$$\int_{a(t)} \epsilon_{ijk}x_j t_k\, da + \int_{v(t)} \epsilon_{ijk}x_j\rho b_k\, dv = \frac{d}{dt}\int_{v(t)} \rho\epsilon_{ijk}x_j v_k\, dv \tag{20.5}$$

As before, the traction can be expressed in terms of the stress as $t_k = n_l \sigma_{lk}$. Using the divergence theorem to rewrite the surface integral as a volume integral and distributing the multiplication yields

$$\int_{a(t)} \epsilon_{ijk} x_j t_k \, \mathrm{d}a = \int_{v(t)} \epsilon_{ijk} \left[\delta_{jl} \sigma_{lk} + x_j \frac{\partial \sigma_{lk}}{\partial x_l} \right] \mathrm{d}v \qquad (20.6)$$

Reynolds' transport theorem can be used to write the right side of (20.5) as

$$\frac{\mathrm{d}}{\mathrm{d}t} \int_{v(t)} \rho \epsilon_{ijk} x_j v_k \, \mathrm{d}v = \int_{v(t)} \rho \epsilon_{ijk} \frac{\mathrm{d}}{\mathrm{d}t} (x_j v_k) \mathrm{d}v \qquad (20.7)$$

where

$$\frac{\mathrm{d}}{\mathrm{d}t} (x_j v_k) = v_j v_k + x_j \frac{\mathrm{d}v_k}{\mathrm{d}t} \qquad (20.8)$$

Using the results of (20.6) to (20.8) in (20.5) yields

$$\int_{v(t)} \epsilon_{ijk} \sigma_{jk} \, \mathrm{d}v + \int_{v(t)} \epsilon_{ijk} x_j \left\{ \frac{\partial \sigma_{lk}}{\partial x_l} + \rho b_k - \rho \frac{\mathrm{d}v_k}{\mathrm{d}t} \right\} \mathrm{d}v - \int_{v(t)} \rho \epsilon_{ijk} v_j v_k \, \mathrm{d}v = 0$$

The last term vanishes because ϵ_{ijk} is skew-symmetric in jk and $v_j v_k$ is symmetric, and the term $\{\ldots\}$ vanishes because of the equation of motion (20.4). Because the remaining integral must vanish for all material volumes v, the integrand must be zero

$$\epsilon_{ijk} \sigma_{jk} = 0$$

Multiplying by ϵ_{ipq}, summing, and using the ϵ–δ identity (4.13) gives

$$\sigma_{pq} = \sigma_{qp} \qquad (20.9)$$

or

$$\sigma = \sigma^T \qquad (20.10)$$

This is a more systematic demonstration of the symmetry of the stress tensor than given in Section 9.4.

20.2   Momentum Balance in the Reference State

In the preceding section, we expressed the balances of linear and angular momenta in terms of integrals over the body in the current configuration. The stress tensor that entered is the Cauchy stress σ discussed in Chapter 9. Often, however, it is more convenient to use the reference configuration. Referring the balance of momentum to the reference state introduces

a different stress tensor. This stress tensor is not symmetric but nevertheless is consistent with
(20.9) and (20.10).

20.2.1 Linear Momentum

Let $\mathbf{t}^0$ be the surface force per unit *reference* area, $\mathbf{b}^0$ be the body force per unit mass, and ρ_0
be the mass density in the reference state. Then application of (20.1) to a volume V enclosed
by a surface A gives

$$\int_A \mathbf{t}^0 \, dA + \int_V \rho_0 \mathbf{b}^0 \, dV = \frac{\partial}{\partial t} \int_V \rho_0 \mathbf{v} \, dV \tag{20.11}$$

Note that $\mathbf{t}^0$ and $\rho_0 \mathbf{b}^0$ express the *current* surface and body force although they are computed
in terms of the *reference* area and volume. Also, the partial derivative, rather than the material
derivative, is used on the right side because the reference volume is not changing in time.
All the quantities in this equation should be considered functions of position in the reference
configuration $\mathbf{X}$. The nominal traction can be written in terms of a stress as

$$\mathbf{t}^0 = \mathbf{N} \cdot \mathbf{T}^0 \tag{20.12}$$

where $\mathbf{N}$ is the unit normal to area A in the reference configuration and $\mathbf{T}^0$ is the nominal
stress (this stress or its transpose is often called the first Piola–Kirchhoff stress) rather than the
Cauchy stress σ. Application of the divergence theorem (20.3) in the reference configuration
to the first term yields

$$\int_A \mathbf{N} \cdot \mathbf{T}^0 \, dA = \int_V \nabla_{\mathbf{X}} \cdot \mathbf{T}^0 \, dV$$

where the subscript $\mathbf{X}$ emphasizes that the derivatives in the divergence are with respect
to position in the reference configuration. Substituting into (20.11) and bringing the partial
derivative inside the integral gives

$$\int_V \left\{ \nabla_{\mathbf{X}} \cdot \mathbf{T}^0 + \rho_0 \mathbf{b}^0 - \rho_0 \frac{\partial \mathbf{v}}{\partial t} \right\} dV = 0$$

Since this integral must vanish for any material volume, the integrand must vanish:

$$\nabla_{\mathbf{X}} \cdot \mathbf{T}^0 + \rho_0 \mathbf{b}^0 = \rho_0 \frac{\partial \mathbf{v}}{\partial t} \tag{20.13}$$

or, in component form,

$$\frac{\partial T^0_{ij}}{\partial X_i} + \rho_0 b^0_j = \rho_0 \frac{\partial v_j}{\partial t}$$

The connection between the Cauchy stress $\boldsymbol{\sigma}$ and the nominal stress $\mathbf{T}^0$ can be established by noting that both must give the same increment of current force $d\mathbf{P}$

$$d\mathbf{P} = \mathbf{n} \cdot \boldsymbol{\sigma} \, da = \mathbf{N} \cdot \mathbf{T}^0 \, dA$$

Using Nanson's formula (15.12) relating the current and reference area elements and rearranging yields the following relation between the nominal and Cauchy stress tensors:

$$\mathbf{T}^0 = \det(\mathbf{F})\mathbf{F}^{-1} \cdot \boldsymbol{\sigma} \tag{20.14}$$

20.2.2 Angular Momentum

The balance of angular momentum can also be expressed in terms of the reference area and volume:

$$\int_A \mathbf{x} \times \mathbf{t}^0 \, dA + \int_V \mathbf{x} \times \rho_0 \mathbf{b}^0 \, dV = \frac{\partial}{\partial t} \int_V \mathbf{x} \times \rho_0 \mathbf{v} \, dV$$

or in component form

$$\int_A \epsilon_{ijk} x_j t_k^0 \, dA + \int_V \epsilon_{ijk} x_j \rho_0 b_k^0 \, dV = \frac{\partial}{\partial t} \int_V \rho_0 \epsilon_{ijk} x_j v_k \, dV \tag{20.15}$$

Note that $\mathbf{x}$, not $\mathbf{X}$, appears in these expressions because the current moment and angular momentum are the cross product of the current location with the current force and linear momentum even though these are expressed in terms of integrals over the reference area and volume. As before, the traction can be expressed in terms of the stress as in (20.12), and using the divergence theorem to rewrite the surface integral as a volume integral gives

$$\int_A \epsilon_{ijk} x_j t_k^0 \, dA = \int_V \epsilon_{ijk} \frac{\partial}{\partial X_l} \left\{ x_j T_{lk}^0 \right\} dV$$

In contrast to the derivation in terms of the current configuration, distributing the derivative in the first term becomes

$$\frac{\partial x_j}{\partial X_l} = F_{jl}$$

rather than δ_{jl}. Because the integral on the right side of (20.15) is over the reference volume, the derivative can be taken inside without recourse to Reynolds' transport theorem. When the balance of linear momentum (20.13) is used, the only term remaining is

$$\int_V \epsilon_{ijk} F_{jl} T_{lk}^0 \, dV = 0$$

Because the integral must vanish for all volumes V, the integrand must be zero

$$\epsilon_{ijk}F_{jl}T^0_{lk} = 0$$

which requires that

$$\mathbf{F} \cdot \mathbf{T}^0 = \left(\mathbf{F} \cdot \mathbf{T}^0\right)^T \tag{20.16}$$

Because the deformation gradient $\mathbf{F}$ is not, in general, symmetric, the nominal stress will not be symmetric. But since the nominal and Cauchy stress are related by (20.10), equation (20.16) is equivalent to the requirement that the Cauchy stress be symmetric.

20.3   Momentum Balance for a Mixture

For a mixture of α constituents, the balance of linear momentum for each constituent is given by

$$\rho_\alpha \left(\frac{\partial \mathbf{v}_\alpha}{\partial t} + \mathbf{v}_\alpha \cdot \nabla \mathbf{v}_\alpha\right) = \nabla \cdot \sigma_\alpha + \mathbf{m}_\alpha + \rho_\alpha \mathbf{b}_\alpha$$

where $\mathbf{v}_\alpha$ is the velocity of the α constituent, σ_α and $\mathbf{b}_\alpha$ are the stress and body force on the α constituent, $\mathbf{m}_\alpha$ is a momentum exchange term, and ρ_α is the mass of α per volume of the mixture, equal to $\phi_\alpha \gamma_\alpha$, where ϕ_α is the volume fraction of α and γ_α is the mass density of α. Summing over all constituents yields

$$\sum_\alpha \rho_\alpha \frac{\partial \mathbf{v}_\alpha}{\partial t} + \sum_\alpha \rho_\alpha \mathbf{v}_\alpha \cdot \nabla \mathbf{v}_\alpha = \sum_\alpha \nabla \cdot \sigma_\alpha + \rho \mathbf{b} \tag{20.17}$$

where $\rho\mathbf{b} = \sum_\alpha \rho_\alpha \mathbf{b}_\alpha$ and $\sum_\alpha \mathbf{m}_\alpha = 0$ since the total momentum exchange for all constituents is zero. The first term on the left can be written as

$$\sum_\alpha \rho_\alpha \frac{\partial \mathbf{v}_\alpha}{\partial t} = \sum_\alpha \frac{\partial(\rho_\alpha \mathbf{v}_\alpha)}{\partial t} - \sum_\alpha \mathbf{v}_\alpha \frac{\partial \rho_\alpha}{\partial t}$$

Using mass conservation for the α constituent

$$\frac{\partial \rho_\alpha}{\partial t} + \nabla \cdot (\rho_\alpha \mathbf{v}_\alpha) = 0$$

to rewrite the second term and substituting back into (20.17) yields

$$\frac{\partial(\rho\mathbf{v})}{\partial t} + \sum_\alpha \left\{\mathbf{v}_\alpha \cdot \nabla \rho_\alpha \mathbf{v}_\alpha + \rho_\alpha \mathbf{v}_\alpha \cdot \nabla \mathbf{v}_\alpha\right\} = \sum_\alpha \nabla \cdot \sigma_\alpha + \rho \mathbf{b}$$

where $\rho \mathbf{v} = \sum_\alpha \rho_\alpha \mathbf{v}_\alpha$. Expanding the first term and using conservation of mass for the summed constituents and the material derivative of $\mathbf{v}$ gives

$$\rho \frac{d}{dt}\mathbf{v} = \nabla \cdot \sum_\alpha (\sigma_\alpha + [\rho \mathbf{v}\mathbf{v} - \rho_\alpha \mathbf{v}_\alpha \mathbf{v}_\alpha]) + \rho \mathbf{b}$$

Note that if the total stress is defined as the term in parentheses on the right side, there is a contribution from the differences of the momentum of the constituents from the average momentum.

Exercises

20.1 The stress tensor is given by

$$\sigma = T\mathbf{n}\mathbf{n}$$

where $\mathbf{n}$ is a unit vector and $T > 0$. If $T = T(\mathbf{x})$, the body is in equilibrium, and is not subject to body forces, show that the gradient of T is perpendicular to $\mathbf{n}$.

20.2 The stress state in a body occupying $|x_1| \le a, |x_2| \le a$ is given by

$$\sigma_{11} = -p \frac{\left(x_1^2 - x_2^2\right)}{a^2}$$

$$\sigma_{22} = p \frac{\left(x_1^2 - x_2^2\right)}{a^2}$$

$$\sigma_{12} = 2px_1 x_2 / a^2$$

and $\sigma_{33} = \sigma_{13} = \sigma_{23} = 0$. If the body is not subjected to body forces, determine whether it is in equilibrium.

20.3 Show that the right side of (20.4) can be written as

$$\rho \frac{dv_j}{dt} = \frac{\partial}{\partial t}\left(\rho v_j\right) + \frac{\partial}{\partial x_i}\left(\rho v_i v_j\right)$$

and use this result to derive the equation for the balance of linear momentum for a control volume fixed in space.

21

Conservation of Energy

The equation expressing conservation of energy for a continuum results from application of the first law of thermodynamics. The first law states that the change in total energy of a system is equal to the sum of the work done on the system and the heat added to the system. Thus, in rate form the first law is

$$\dot{E}_{\text{total}} = P + \dot{Q} \tag{21.1}$$

where P is the power input and $\dot{Q}$ is the rate of heat input. Although neither heat nor work is an exact differential (does not integrate to a potential function), their sum is. Consequently, the integral of the energy change around a cycle is zero. The total energy is the sum of the kinetic energy

$$\text{KE} = \int_{v(t)} \frac{1}{2} \rho \mathbf{v} \cdot \mathbf{v} \, dv$$

where $\mathbf{v}$ is the velocity, and the internal energy

$$\text{IE} = \int_{v(t)} \rho u \, dv$$

where u is the internal energy per unit mass. The rate of heat input is

$$\dot{Q} = -\int_{a(t)} \mathbf{q} \cdot \mathbf{n} \, da + \int_{v(t)} \rho r \, dv$$

where $\mathbf{q}$ is the heat flux, $\mathbf{n}$ is the outward normal to the surface a(t), and r is the rate of internal heating (heat source) per unit mass.

Fundamentals of Continuum Mechanics, First Edition. John W. Rudnicki.
© 2015 John Wiley & Sons, Ltd. Published 2015 by John Wiley & Sons, Ltd.

The power input is the work of the forces on the velocities

$$P = \int_{a(t)} \mathbf{t} \cdot \mathbf{v} \, da + \int_{v(t)} \rho \mathbf{b} \cdot \mathbf{v} \, dv$$

where $\mathbf{t}$ is the traction and $\mathbf{b}$ is the body force per unit mass. Expressing the traction in terms of the stress and using the divergence theorem (18.2) yields the following for the first term:

$$\int_{a(t)} \mathbf{t} \cdot \mathbf{v} \, da = \int_{a(t)} \mathbf{n} \cdot \boldsymbol{\sigma} \cdot \mathbf{v} \, da = \int_{v(t)} \nabla \cdot (\boldsymbol{\sigma} \cdot \mathbf{v}) \, dv \qquad (21.2)$$

To work out $\nabla \cdot (\boldsymbol{\sigma} \cdot \mathbf{v})$ it is convenient to use index notation (see Problem 21.1). After converting back to coordinate-free form, the result is

$$\nabla \cdot (\boldsymbol{\sigma} \cdot \mathbf{v}) = (\nabla \cdot \boldsymbol{\sigma}) \cdot \mathbf{v} + \boldsymbol{\sigma} \cdot \cdot \mathbf{L} \qquad (21.3)$$

where $\mathbf{L}$ is the velocity gradient tensor (14.3). Using the equation of motion (20.4) to rewrite the first term on the right in (21.3) and substituting back into (21.2) yields

$$P = \int_{v(t)} \rho \frac{1}{2} \frac{d}{dt} (\mathbf{v} \cdot \mathbf{v}) \, dv + \int_{v(t)} \boldsymbol{\sigma} \cdot \cdot \mathbf{L} \, dv \qquad (21.4)$$

Using Reynolds' transport theorem on the first term gives

$$\int_{v(t)} \frac{1}{2} \rho \frac{d}{dt} (\mathbf{v} \cdot \mathbf{v}) \, dv = \frac{d}{dt} \int_{v(t)} \frac{1}{2} \rho \mathbf{v} \cdot \mathbf{v} \, dv \qquad (21.5)$$

Substituting (21.5) into (21.4) and then (21.4) and (21.3) back into (21.1) and cancelling the common term, $d(KE)/dt$, on each side yields

$$\frac{d}{dt} \int_{v(t)} \rho u \, dv = \int_{v(t)} \boldsymbol{\sigma} \cdot \cdot \mathbf{L} \, dv - \int_{a(t)} \mathbf{q} \cdot \mathbf{n} \, da + \int_{v(t)} \rho r \, dv$$

The divergence theorem (18.1) can be used to convert the heat flux term to a volume integral. Using Reynolds' transport theorem on the internal energy term and collecting terms gives

$$\int_{v(t)} \left\{ \rho \frac{du}{dt} - \boldsymbol{\sigma} \cdot \cdot \mathbf{L} + \nabla \cdot \mathbf{q} - \rho r \right\} dv = 0$$

Since this applies for all v the local form of energy conservation is

$$\rho \frac{du}{dt} = \boldsymbol{\sigma} \cdot \cdot \mathbf{L} - \nabla \cdot \mathbf{q} + \rho r \qquad (21.6)$$

This equation has the simple interpretation that the internal energy of a continuum can be changed by the work of deformation, $\boldsymbol{\sigma} \cdot \cdot \mathbf{L}$, the flow of heat, $-\nabla \cdot \mathbf{q}$, or internal heating, ρr.

Similarly to the momentum balance equations, the energy equation can be expressed in terms of quantities per unit area and volume of the reference configuration. Manipulations similar to those above lead to

$$\rho_0 \frac{\partial u}{\partial t} = \mathbf{T}^0 \cdot\cdot \dot{\mathbf{F}} - \nabla_{\mathbf{X}} \cdot \mathbf{Q} + \rho_0 R \qquad (21.7)$$

where $\mathbf{Q}$ is the heat flux per unit reference area

$$\mathbf{Q} = J\mathbf{F}^{-1} \cdot \mathbf{q}$$

$\rho_0 R$ is the rate of internal heating per unit reference volume and $\nabla_{\mathbf{X}}(...)$ denotes the gradient with respect to position in the reference configuration.

21.1 Work-Conjugate Stresses

The first term on the right side of (21.7) is the rate of stress working per unit reference volume:

$$\dot{W}_0 = \mathbf{T}^0 \cdot\cdot \dot{\mathbf{F}} \qquad (21.8)$$

Since the first term on the right side of (21.6) is the rate of stress working per unit current volume, it is related to (21.8) by

$$\dot{W}_0 = J\boldsymbol{\sigma} \cdot\cdot \mathbf{L} = J\boldsymbol{\sigma} \cdot\cdot D \qquad (21.9)$$

where $J = \det(\mathbf{F})$ and the second equality makes use of the symmetry of $\boldsymbol{\sigma}$. The relation between the Cauchy stress $\boldsymbol{\sigma}$ and the nominal stress $\mathbf{T}^0$ (20.12) can be obtained by equating the two expressions (21.8) and (21.9)

$$\mathbf{T}^0 \cdot\cdot \dot{\mathbf{F}} = J\boldsymbol{\sigma} \cdot\cdot \mathbf{L}$$

Substituting $\dot{\mathbf{F}} = \mathbf{L} \cdot \mathbf{F}$ in the left side gives

$$\mathbf{T}^0 \cdot\cdot (\mathbf{L} \cdot \mathbf{F}) = J\boldsymbol{\sigma} \cdot\cdot \mathbf{L} \qquad (21.10)$$

The identity of Problem 3.9 can be used to rearrange (21.10) as

$$\left(\mathbf{F} \cdot \mathbf{T}^0 - J\boldsymbol{\sigma} \right) \cdot\cdot \mathbf{L} = 0$$

Since this must apply for any velocity gradient tensor $\mathbf{L}$, the coefficient must vanish and, therefore, the nominal stress is given by the same relation derived from Nanson's formula (21.10) for the current and reference areas:

$$\mathbf{T}^0 = J\mathbf{F}^{-1} \cdot \boldsymbol{\sigma}$$

In both (21.8) and (21.9) $\dot{W}_0$ is the product of a stress tensor and a deformation rate measure. The stress measure is said to be *work conjugate* to the rate of deformation measure. Note that the stress measure work conjugate to $\mathbf{L}$ or $\mathbf{D}$ is not the Cauchy stress but the Kirchhoff stress, which is the product

$$\tau = J\sigma$$

This distinction, although small if volume changes are small, can be important in numerical formulations. Even though σ and $\mathbf{D}$ are both symmetric, the stiffness matrix in a finite element formulation is not guaranteed to be symmetric only if based on a relation in terms of $\mathbf{D}$ and σ.

More generally, the relation for the rate of stress working per unit reference volume can be used to define symmetric stress tensors $\mathbf{S}$ that are work conjugate to the rate of any material strain tensor $\dot{\mathbf{E}}$. (Since the rate of a material strain tensor is symmetric, there is no point in retaining any antisymmetric part to the conjugate stress tensor since it does not contribute to $\dot{W}_0$.) Thus, writing

$$\dot{W}_0 = \mathbf{S} \cdot\cdot \dot{\mathbf{E}} \tag{21.11}$$

and equating to (21.8) or (21.9) defines $\mathbf{S}$ for a particular rate of material strain $\dot{\mathbf{E}}$. For example, we can determine the stress measure that is work conjugate to the rate of Green–Lagrange strain $\dot{\mathbf{E}}^G$

$$\dot{W}_0 = J\sigma \cdot\cdot \mathbf{D} = \mathbf{S}^{PK2} \cdot\cdot \dot{\mathbf{E}}^G$$

Using the relation between the rate of Green–Lagrange strain and the rate of deformation (16.16) yields

$$\left(J\sigma - \mathbf{F} \cdot \mathbf{S}^{PK2} \cdot \mathbf{F}^T\right) \cdot\cdot \mathbf{D} = 0$$

Since this must apply for any $\mathbf{D}$, the work-conjugate stress is given by

$$\mathbf{S}^{PK2} = \mathbf{F}^{-1} \cdot J\sigma \cdot (\mathbf{F}^T)^{-1} \tag{21.12}$$

and it is clearly symmetric. This stress measure is called the second Piola–Kirchhoff stress.

The second Piola–Kirchhoff stress has the advantages that it is symmetric and that it is work conjugate to the rate of the Green-Lagrange strain, the most commonly used finite strain tensor. It does, however, have the disadvantage that its interpretation in terms of a force increment is less straightforward than either the Cauchy stress σ or the nominal stress $\mathbf{T}^0$. The force increment is related to the traction vector determined from $\mathbf{S}^{PK2}$ by

$$\mathbf{N}\,dA \cdot \mathbf{S}^{PK2} = \mathbf{F}^{-1} \cdot d\mathbf{P}$$

Thus, the traction derived from $\mathbf{S}^{PK2}$ is related to the force per reference area but altered by $\mathbf{F}^{-1}$. The components of this traction do have a direct interpretation in terms of force components expressed in terms of base vectors that convect (deform with the material).

Exercises

21.1 Use index notation to verify (21.3).

21.2 If the stress is given by $\sigma = -p\mathbf{I}$ show that the energy equation can be written as

$$\rho \frac{du}{dt} = \frac{p}{\rho} \frac{d\rho}{dt} - \nabla \cdot \mathbf{q} + \rho r$$

21.3 If the material velocity is zero, the internal energy is a function of temperature θ, and the heat flux is given by Fourier's law

$$\mathbf{q} = -\kappa \nabla \theta$$

with κ constant, show that the energy equation reduces to the usual form of the heat equation

$$\frac{\partial \theta}{\partial t} = \alpha \nabla^2 \theta + r/c$$

where $\alpha = \kappa/\rho c$ is a diffusivity, $c = \partial u/\partial \theta$ is a specific heat, and $\nabla \cdot \nabla (\ldots) = \nabla^2 (\ldots)$.

21.4 Beginning with a statement of energy balance in terms of integrals over the reference configuration, derive (21.7). Clearly define the terms entering this equation and comment on their relation to the corresponding terms in (21.6).

21.5 Show that the stress work conjugate to the material strain tensor $\mathbf{E}^{(-2)}$ is

$$\mathbf{S}^{(-2)} = \mathbf{F}^T \cdot J\sigma \cdot \mathbf{F}$$

21.6 Show that the stress work conjugate to the rate of the stretch measure of strain $\mathbf{E}^{(1)} = \mathbf{U} - \mathbf{I}$ is

$$\mathbf{S}^{(1)} = \frac{1}{2} J \{ \mathbf{U}^{-1} \cdot (\mathbf{R}^T \cdot \sigma \cdot \mathbf{R}) + (\mathbf{R}^T \cdot \sigma \cdot \mathbf{R}) \cdot \mathbf{U}^{-1T} \}$$

Part Five

Ideal Constitutive Relations

Thus far, we have analyzed stress, strain, rate of deformation, and the laws expressing conservation of mass, momentum, and energy. Nowhere have we incorporated the behavior of particular materials. This is a large and complex subject. Inevitably, descriptions of material behavior are idealized relationships between stress and strain or rate of deformation and their history. Ultimately, such relationships derive from experiments, but they generally apply only for a limited range of states, i.e., temperature, loading rate, time-scale, etc. Increases in computational power have made it possible to consider material behaviors that are far more complex than in the past. Consequently, the topic of constitutive relations is an enormous one. Here we give a minimal discussion of the simplest idealizations.

Crudely, materials can be divided into solids (which sustain shear stress at rest) and fluids (which cannot), but many materials combine aspects of both. Chapter 22 discusses simple fluid idealizations and Chapter 23 discusses elastic idealizations of solids.

Fundamentals of Continuum Mechanics, First Edition. John W. Rudnicki.
© 2015 John Wiley & Sons, Ltd. Published 2015 by John Wiley & Sons, Ltd.

22

Fluids

22.1   Ideal Frictionless Fluid

Observations indicate that a fluid at rest or in uniform motion cannot support shear stress. Consequently, the stress is given by

$$\sigma_{ij} = -p\delta_{ij} \tag{22.1}$$

where p is a pressure. If the fluid is at rest or if local thermodynamic equilibrium is assumed, then p is the thermodynamic pressure (Aris, 1989, Sec. 5.14). If the stress is assumed to have this form regardless of the motion, the fluid is *ideal* or *perfect*. Since p is an unknown, another equation is needed to determine it. This can be an equation of state of the form

$$F(p, \rho, \theta) = 0 \tag{22.2}$$

where ρ is the mass density and θ is the temperature. A simple example is the perfect gas law

$$p = \rho R \theta$$

where R is the universal gas constant.

   If temperature does not play a role, the flow is *barotropic* and the pressure is related to the density by an equation of the form

$$f(p, \rho) = 0$$

An equation of state (22.2) reduces to this form for either constant temperature (isothermal) or no heat flow (isentropic). For example, for isentropic flow of a perfect gas

$$\frac{p}{\rho^{\gamma}} = \text{constant}$$

In this equation

$$\gamma = \frac{c_p}{c_v} = 1 + \frac{R}{c_v}$$

Fundamentals of Continuum Mechanics, First Edition. John W. Rudnicki.
© 2015 John Wiley & Sons, Ltd. Published 2015 by John Wiley & Sons, Ltd.

where c_p and c_v are the specific heat at constant pressure and constant volume, respectively. For dry air, $\gamma = 1.4$.

Alternatively, the internal energy per unit mass can be prescribed as a function of the density and the temperature:

$$u = u(\theta, \rho)$$

Recall that conservation of energy is expressed by (21.6)

$$\rho \frac{du}{dt} = \boldsymbol{\sigma} \cdot \cdot \mathbf{D} - \boldsymbol{\nabla} \cdot \mathbf{q} + \rho r$$

where r is the rate of internal heating per unit mass, $\mathbf{q}$ is the flux of heat (out of the body), and $\mathbf{D}$ replaces $\mathbf{L}$ in (21.6) because $\boldsymbol{\sigma}$ is symmetric. Substituting (22.1) and using conservation of mass (19.9) gives

$$\rho \frac{du}{dt} = p \frac{1}{\rho} \frac{d\rho}{dt} - \boldsymbol{\nabla} \cdot \mathbf{q} + \rho r \qquad (22.3)$$

If u is regarded as a function of $v = 1/\rho$, the specific volume (rather than the density), and the temperature θ, then the material derivative of u on the left side of (22.3) can be written as

$$\frac{du}{dt} = \frac{\partial u}{\partial v} \frac{dv}{dt} + c_v \frac{d\theta}{dt} \qquad (22.4)$$

where $c_v = \partial u / \partial \theta$ is the specific heat at constant volume. Substituting (22.4) into (22.3) and rearranging gives

$$\rho c_v \frac{d\theta}{dt} = -\frac{1}{v} \frac{dv}{dt} \left(p + \frac{\partial u}{\partial v} \right) + \rho r - \boldsymbol{\nabla} \cdot \mathbf{q} \qquad (22.5)$$

This equation must hold for all motions of the fluid including those for which the temperature is constant. At constant temperature, all terms, except the first on the right, vanish and (22.5) requires that

$$p = -\frac{\partial u}{\partial v}$$

This equation provides a constitutive relation for the pressure in terms of the dependence of the energy on the specific volume. According to the terminology used in Chapter 21, the pressure and specific volume are work-conjugate variables.

If the material is incompressible so that $d\rho/dt = 0$, then the mechanical response uncouples from the thermal response governed by

$$\rho c_v \frac{d\theta}{dt} = \rho r - \boldsymbol{\nabla} \cdot \mathbf{q} \qquad (22.6)$$

The rate of heating per unit mass r is regarded as prescribed but a constitutive equation is needed to relate the heat flux $\mathbf{q}$ to the temperature. (Considerations based on the second law of thermodynamics, not discussed here, indicate that these are the proper variables to relate.) Typically, this relation is taken to be Fourier's law, which states that the heat flux is proportional to the negative gradient of temperature

$$\mathbf{q} = -\boldsymbol{\kappa} \cdot \nabla\theta \qquad (22.7)$$

or in component form

$$q_i = -\kappa_{ij}\partial\theta/\partial x_j$$

The thermal conductivity tensor $\boldsymbol{\kappa}$ depends on the material. Second law of thermodynamics considerations, again not discussed here, require that $\boldsymbol{\kappa}$ be symmetric, $\boldsymbol{\kappa} = \boldsymbol{\kappa}^T$. If $\boldsymbol{\kappa}$ does not depend on position, then the material is *homogeneous* (with respect to heat conduction). If the material has no directional properties and heat conduction is the same in all directions, then the material is *isotropic*. In this case, $\boldsymbol{\kappa}$ is an isotropic tensor of the form

$$\boldsymbol{\kappa} = k\mathbf{I}$$

(see Section 6.4). Substituting into (22.7) and then (22.6) gives

$$\rho c_v \frac{d\theta}{dt} = \rho r + k\nabla^2\theta \qquad (22.8)$$

if the velocity is small enough so that $d\theta/dt \approx \partial\theta/\partial t$, then (22.8) reduces to the usual form of the heat equation

$$\frac{\partial\theta}{\partial t} = r/c_v + \alpha\nabla^2\theta \qquad (22.9)$$

where $\alpha = k/\rho c_v$ is the thermal diffusivity (with dimensions of length squared per time).

22.2   Linearly Viscous Fluid

In a simple idealization of a fluid, the stress is taken to be the sum of a hydrostatic term and a function of the rate of deformation:

$$\boldsymbol{\sigma} = -p\mathbf{I} + \mathbf{f}(\mathbf{D})$$

where the function $\mathbf{f}$ vanishes when $\mathbf{D} = 0$. Such a fluid is called *Stokesian* (see Aris 1989, Secs. 5.21 and 5.22, pp. 106–109).

If the stress depends linearly on the rate of deformation,

$$\sigma_{ij} = -p\delta_{ij} + V_{ijkl}D_{kl} \qquad (22.10)$$

the fluid is *Newtonian*. In this case the factors V_{ijkl} may depend on temperature but not on stress or deformation rate. Because $\sigma_{ij} = \sigma_{ji}$ and $D_{kl} = D_{lk}$, there are six distinct components of σ_{ij} and D_{ij}, and a total of $36 = 6 \times 6$ possible distinct components of V_{ijkl}. If the V_{ijkl} are assumed to have the additional symmetry $V_{ijkl} = V_{klij}$, the number of possible distinct components is reduced to 21.

If the material response is completely independent of the orientation of axes, the material is *isotropic*. In this case, $\mathbf{V}$ is an isotropic tensor and, as discussed in Chapter 6, has the form (6.16) with $c = 0$ because the coefficient term is antisymmetric with respect to the interchange of (i) and (j). Substituting into (22.10) yields

$$\sigma_{ij} = -p\delta_{ij} + \lambda\delta_{ij}D_{kk} + 2\mu D_{ij} \tag{22.11}$$

where λ and μ are the only two parameters reflecting material response. Taking the trace of (22.11) yields

$$\frac{1}{3}\sigma_{kk} = -p + \Theta D_{kk}$$

where $\Theta = \lambda + 2\mu/3$ is the bulk viscosity. Taking the deviatoric part of (22.11) yields

$$\sigma'_{ij} = 2\mu D'_{ij}$$

where μ is the shear viscosity. Substituting (22.11) into the equation of motion (20.4) gives

$$(\mu + \lambda)\frac{\partial}{\partial x_j}\left(\frac{\partial v_k}{\partial x_k}\right) + \mu\nabla^2 v_j - \frac{\partial p}{\partial x_j} + \rho b_j = \rho\frac{dv_j}{dt} \tag{22.12}$$

or, in coordinate-free form,

$$(\mu + \lambda)\nabla\nabla \cdot \mathbf{v} + \mu\nabla^2\mathbf{v} - \nabla p + \rho\mathbf{b} = \rho\frac{d\mathbf{v}}{dt}$$

These are the Navier–Stokes equations. If the flow is isochoric (involves no volume change), $\partial v_k/\partial x_k = 0$, and (22.12) reduces to

$$\mu\nabla^2 v_j - \frac{\partial p}{\partial x_j} + \rho b_j = \rho\frac{dv_j}{dt}$$

The shear viscosity μ can be determined by a simple conceptual experiment. Consider a layer of fluid of height h between two parallel plates with lateral dimensions much greater than h and no body force or pressure gradient (Figure 22.1). (The geometry discussed here is clearly impractical. In reality the experiment is conducted with an arrangement of concentric rotating cylinders. See Problem 22.2.) The upper plate ($x_2 = h$) is moved to the right (positive x_1 direction) with velocity V. Consequently, the conditions on the fluid velocity at the boundaries are $v_1(x_2 = h) = V$ and $v_1(x_2 = 0) = 0$.

After a transient that occurs immediately after the plate begins moving, the velocity in the fluid does not depend on time, i.e., the flow is steady. Integration and use of the boundary conditions give $v_1 = x_2 V/h$. The only nonzero component of D_{ij} is

$$D_{12} = D_{21} = \frac{1}{2}\left(\frac{\partial v_1}{\partial x_2} + \frac{\partial v_2}{\partial x_1}\right) = \frac{V}{2h} = \frac{1}{2}\dot{\gamma}$$

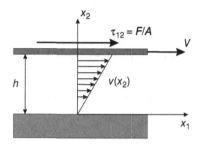

Figure 22.1   Flow between infinite plates. Lower plate $x_2 = 0$ is stationary. Upper plate $x_2 = h$ is moving to the right with velocity V.

and the shear stress σ_{12} is the force applied to the plate divided by its contact area with the fluid. If a plot of σ_{12} against $\dot{\gamma}$ is linear, then the fluid is Newtonian and viscosity is μ. Commonly the viscosity is measured in poise, which is equal to 1 dyne-s/cm^2. In poise, representative viscosities for water, air, and SAE 30 oil are 10^{-2}, 1.8×10^{-4}, and 0.67 respectively.

22.2.1   Non-steady Flow

As an example of non-steady flow, consider a plate large in the x_1 and x_3 directions that bounds a semi-infinite expanse of fluid ($x_2 = y \geq 0$). At time $t = 0$, the plate moves in the positive x_1 direction with velocity V. The only nonzero component of velocity is $v_1 = v(y, t)$. For a velocity of this form and no body force and pressure gradient, (22.2) reduces to

$$\frac{\partial^2 v}{\partial y^2} = \frac{1}{\eta}\frac{\partial v}{\partial t} \tag{22.13}$$

where $\eta = \mu/\rho$ is the dynamic viscosity. In m^2/s, η equals approximately 10^{-6}, 1.5×10^{-5}, and 7.3×10^{-5} for water, air and SAE 30 oil respectively. At $y = 0$ and $t \geq 0$, $v = V$. The fluid velocity must vanish as $y \to \infty$. (Note that (22.13) is identical in form to the heat equation (22.9) without the source term and, hence, the solution described below is also a solution to the corresponding thermal problem.)

To construct a solution, note that there is no characteristic length in the problem. Consequently, the velocity can depend on y only in the non-dimensional form $\xi = y/\sqrt{4\eta t}$, where the factor of 4 is inserted purely for convenience. Because the problem is linear, the velocity must also be proportional to V and, hence, have the form

$$v(y, t) = Vf(\xi) \tag{22.14}$$

where $f(\xi)$ is a function to be determined. Substituting (22.14) into (22.13) yields

$$f''(\xi) + 2\xi f'(\xi) = 0$$

and integrating gives

$$f(\xi) = A \int_0^{\xi} \exp(-s^2)\,ds + B \qquad (22.15)$$

where A and B are constants. Using the condition at $y = 0$ gives $B = 1$. For $y \to \infty$, $\xi \to \infty$ and (22.15) gives

$$A \int_0^{\infty} \exp(-s^2)\,ds + B = 0$$

The value of the integral is $\sqrt{\pi}/2$ and $A = -2/\sqrt{\pi}$. The velocity is given by

$$v(y, t) = Verfc\left(y/\sqrt{4\eta t}\right)$$

where

$$erfc(x) = \frac{2}{\sqrt{\pi}} \int_x^{\infty} \exp(-s^2)\,ds$$

is the complementary error function and the error function $erf(x) = 1 - erfc(x)$. Figure 22.2 plots both the complementary error function and the error function.

Because $erfc(2) = 0.005$, the fluid velocity is reduced to 0.5% of the plate velocity. At 10 s, this occurs at 29, 13 and 108 mm from the plate for air, water, and SAE 30 oil, indicating that the effect of viscosity is confined to the neighborhood of the plate. This observation is the foundation of boundary layer theory in which the viscosity is included only near a boundary and the velocity there is matched to the flow of an inviscid fluid away from the boundary.

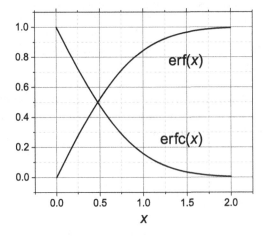

Figure 22.2 Error function and complementary error function.

Exercises

22.1 Consider Poiseuille flow between two large (essentially unbounded) plates separated by a distance $2h$ (Figure 22.3). The plates are stationary, the flow is steady $(\partial(\ldots)/\partial t = 0)$, and there are no body forces. Velocity is only in the x_1 direction and depends only on x_2; that is, $v_1 = v(x_2)$ is the only non-zero velocity component. The flow occurs in response to a constant pressure gradient in the x_1 direction, dp/dx_1.

    (a) Determine the velocity profile.

    (b) Determine the shear stress σ_{12} at $x_2 = h$.

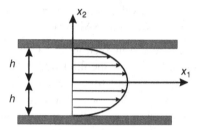

Figure 22.3   Plane Poiseuille flow between two parallel plates.

22.2 As mentioned, a practical arrangement for determining the shear viscosity μ is flow between concentric cylinders, as depicted in Figure 22.4. The outer cylinder of radius b is stationary and the inner cylinder of radius a is rotating with angular velocity Ω. The cylinders are long in the out-of-plane direction and the flow is steady (the cylinder has been rotating for a long time so that $\partial(\ldots)/\partial t = 0$). Flow is only in the circumferential direction and depends only on radial distance, i.e., $\mathbf{v} = v(r)\mathbf{e}_\theta$. There are no body forces or pressure gradient. [Hint: The problem can be solved using the answers to Problems 8.11 to 8.14. The components of the rate of deformation $\mathbf{D}$ are given by the symmetric part of the answer to Problem 8.13 and the only nonzero component is

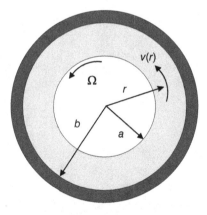

Figure 22.4   Cylindrical Couette flow between a rotating inner cylinder of radius a and a stationary concentric outer cylinder of radius b.

$D_{r\theta} = 2\mu\sigma_{r\theta}$. Using the answer to Problem 8.14 to determine the equilibrium equation yields an equation for $v(r)$. Alternatively, the equation for $v(r)$ can be obtained by applying the answer to Problem 8.11 to $v(r)\mathbf{e}_\theta$, taking account of the derivatives of $\mathbf{e}_\theta$.]

(a) Determine the velocity profile between the cylinders.

(b) Determine the shear stress on the inner cylinder.

(c) Determine the relation between the torque and angular velocity on the inner cylinder and explain how it can be used to determine the viscosity.

22.3 Consider the same problem solved in Section 22.2.1, but now the plate is stationary and the velocity far away from the plate ($y \to \infty$) is V for all $t \geq 0$. Determine the solution for the velocity and determine the velocity (as a percentage of V) at a distance of 1 mm from the plate after 1 s for air, water and SAE 30 oil.

22.4 Consider the same problem solved in Section 22.2.1, but at $t = 0$ the plate begins to oscillate as $V = V_0 \cos(\omega t)$. Determine the fluid velocity as a function of y and t after the plate has been oscillating for a long time. Plot the velocity (divided by V_0) against the non-dimensional distance $\sqrt{\omega/2c}\, y$ for several values of the non-dimensional time ωt. [Hint: You may find it easier to work with $V = V_0 \exp(i\omega t)$ and take the real part.]

Reference

Aris R 1989 *Vectors, Tensors, and the Basic Equations of Fluid Mechanics*. Dover.

23

Elasticity

23.1 Nonlinear Elasticity

The simple fluid constitutive relations we considered in the last chapter depended only on the rate of deformation (rather than the strain) and, hence, the issue of the appropriate large-strain measure does not arise. The response of solids does, in general, depend on the strain. Fortunately, for many applications, the magnitude of the strain is small, and this makes it possible to consider a linearized problem that introduces considerable simplification. Although this is often a very good approximation, it should be noted that it is strictly valid only for infinitesimal displacement gradients and needs to be reevaluated whenever this is not the case. Before specializing to the case of linearized elasticity, we consider some more general descriptions of elastic materials for finite strain.

23.1.1 Cauchy Elasticity

A minimal definition of an elastic material is one for which the stress depends only on the deformation gradient (rather than, say, the deformation history, or various internal variables)

$$\boldsymbol{\sigma} = \mathbf{g}(\mathbf{F}) \tag{23.1}$$

This formulation is typically referred to as *Cauchy elasticity*. Other features often associated with elasticity are the existence of a strain energy function, a one-to-one relation between stress and strain measures, deformation does not result in any energy loss, or the body recovers its initial shape upon unloading.

Since the relation (23.1) reflects material behavior we expect it to be independent of rigid body rotations. This is called the *principle of frame indifference* or *material objectivity*. A consequence is that the relation (23.1) should depend only on the deformation $\mathbf{U}$ and not the rotation $\mathbf{R}$ in the polar decomposition $\mathbf{F} = \mathbf{R} \cdot \mathbf{U}$. If we consider a pure deformation $\mathbf{U}$, then (23.1) becomes

$$\boldsymbol{\sigma} = \sigma_{KL}\mathbf{N}_K\mathbf{N}_L = \mathbf{g}(\mathbf{U}) \tag{23.2}$$

Fundamentals of Continuum Mechanics, First Edition. John W. Rudnicki.
© 2015 John Wiley & Sons, Ltd. Published 2015 by John Wiley & Sons, Ltd.

where σ_{KL} are components of the Cauchy stress with respect to the principal axes of $\mathbf{U}$ in the reference state. Application of a rotation $\mathbf{R}$ causes only a rotation but no stretching or additional stress so that

$$\sigma = \sigma_{KL}\mathbf{n}_K\mathbf{n}_L \tag{23.3}$$

That is, the components of σ do not change but are now with respect to the principal axes in the current state. Since $\mathbf{n}_K = \mathbf{R} \cdot \mathbf{N}_K = \mathbf{N}_K \cdot \mathbf{R}^T$, (23.3) becomes

$$\sigma = \mathbf{R} \cdot (\sigma_{KL}\mathbf{N}_K\mathbf{N}_L) \cdot \mathbf{R}^T$$

or, using (23.2),

$$\sigma = \mathbf{R} \cdot \mathbf{g}(\mathbf{U}) \cdot \mathbf{R}^T$$

The result can be rewritten as

$$\mathbf{R}^T \cdot \sigma \cdot \mathbf{R} = \mathbf{g}(\mathbf{U}) \tag{23.4}$$

The quantity on the left side is the rotationally invariant Cauchy stress $\hat{\sigma}$. Independence of the constitutive relation to rigid body rotations requires that $\hat{\sigma}$ be a function of the deformation $\mathbf{U}$. Because $\mathbf{U}$ and $\mathbf{R}$ are not easily computed, it is more convenient to rewrite (23.4) by defining

$$\mathbf{g}(\mathbf{U}) = \mathbf{U} \cdot \mathbf{h}(\mathbf{U}^2) \cdot \mathbf{U}^T \tag{23.5}$$

Substituting (23.5) into (23.4), multiplying from the right by $\mathbf{R}$ and from the left by $\mathbf{R}^T$, and noting that $\mathbf{F}^T \cdot \mathbf{F} = \mathbf{U}^2$ gives

$$\sigma = \mathbf{F} \cdot \mathbf{h}(\mathbf{F}^T \cdot \mathbf{F}) \cdot \mathbf{F}^T$$

Further rearrangement gives

$$\mathbf{S}^{PK2} = \mathbf{k}(\mathbf{E}^G) \tag{23.6}$$

where $\mathbf{S}^{PK2}$ is the second Piola–Kirchhoff stress (21.12) and $\mathbf{E}^G$ is the Green–Lagrange strain (16.3), (16.4), (16.5), or (16.6). Thus, a constitutive relation in this form is guaranteed to be independent of rigid body rotations. More generally, this is true for a similar relation between any material strain measure $\mathbf{E}$ and the corresponding work-conjugate stress tensor $\mathbf{S}$.

23.1.2 Green Elasticity

Green elasticity assumes the existence of a strain energy density function W. The existence of W can be motivated by the conservation of energy (21.7):

$$\frac{\partial W}{\partial t} = \mathbf{S} \cdot \cdot \dot{\mathbf{E}} - \nabla_X \cdot \mathbf{Q} + \rho_0 R \tag{23.7}$$

here written in the reference state where $\mathbf{S}$ and $\dot{\mathbf{E}}$ are work-conjugate stress and strain-rate measures (see (21.11)) and $W = \rho_0 u$ is the internal energy per unit reference volume. For isothermal (constant temperature) or adiabatic (no heat flow) conditions the last two terms are absent. Regarding W as a function of $\mathbf{E}$ leads to

$$\left(\frac{\partial W}{\partial E_{ij}} - S_{ij} \right) \dot{E}_{ij} = 0$$

Because this must apply for all $\dot{E}_{ij}$,

$$S_{ij} = \frac{\partial W}{\partial E_{ij}} \tag{23.8}$$

where W is to be written symmetrically in E_{ij} and E_{ji}. If W is regarded as a function of the deformation gradient $\mathbf{F}$, then similar considerations based on (23.7) yield the following expression for the nominal stress:

$$T_{ij}^0 = \frac{\partial W}{\partial F_{ji}} \tag{23.9}$$

23.1.3 Elasticity of Pre-stressed Bodies

As already noted, the strains are small for many practical applications. Often, the elasticity equations for this idealization are stated directly without reference to a more general formulation for arbitrary deformation magnitudes. Seeing how these equations arise from linearization of a more general formulation is, however, an educational exercise. Moreover, we will see that if the material response is linearized about a pre-stressed state, then it is not sufficient for the strains to be small to reduce the formulation to the usual one of linear elasticity. More specifically, the moduli governing changes in the different stress measures will depend on the pre-stress and, consequently, it is necessary to retain the distinction between the different stress measures.

The stress–strain relation is given by (23.6) or by (23.8) if a strain energy function exists. Here we specialize to the case of Green–Lagrange strain and the work-conjugate stress measure, the second Piola–Kirchhoff stress. To simplify the notation, we drop the superscripts $PK2$ and G. Now, we expand the stress–strain relation in a Taylor series about the strain-free state:

$$S_{ij} = (S_{ij})_{\mathbf{E}=0} + C_{ijkl} E_{kl} + B_{ijklmn} E_{kl} E_{mn} + \ldots \tag{23.10}$$

Since deformation is measured from the reference state

$$(S_{ij})_{\mathbf{E}=0} = \bar{\sigma}_{ij}$$

is the Cauchy stress in the reference state. Because the displacement gradients are assumed to be small, the quadratic terms in Green–Lagrange strain can be neglected, E_{ij} reduces to the infinitesimal strain tensor ε_{ij}, (17.6) and (17.7), and only linear terms need be retained in (23.10). Therefore, the stress is given by

$$S_{ij} = \bar{\sigma}_{ij} + C_{ijkl} \varepsilon_{kl} \tag{23.11}$$

where the neglected terms are at least as small as $|\partial u_i / \partial X_j|^2$.

Changes in the second Piola–Kirchhoff stress $\delta S_{ij} = S_{ij} - \bar{\sigma}_{ij}$ are given by

$$\delta S_{ij} = C_{ijkl} \epsilon_{kl}$$

The moduli C_{ijkl} are symmetric with respect to the interchange of indices i and j and k and l because of the symmetry of the stress and the strain. As a consequence of the former, the strain ϵ_{kl} can be replaced by the displacement gradient $u_{k,l} = \partial u_k / \partial X_l$. If a strain energy density function exists (23.8), then the moduli have the additional symmetry

$$C_{ijkl} = \frac{\partial^2 W}{\partial \epsilon_{ij} \partial \epsilon_{kl}} = C_{klij} \tag{23.12}$$

because the derivatives can be taken in either order. In the following, we will assume that a strain energy density function exists.

Because the derivatives entering the Green–Lagrange strain are with respect to position in the reference configuration, we use the equation of motion referred to the reference state:

$$\frac{\partial T_{ij}^0}{\partial X_i} + \rho_0 b_j^0 = \rho_0 \frac{\partial^2 u_j}{\partial t^2} \tag{23.13}$$

where all quantities are to be thought of as functions of position in the reference configuration and time. The nominal stress is related to the nominal traction by

$$N_i T_{ij}^0 = t_j^0 \tag{23.14}$$

on the boundary of the body. Consequently, we need to express the nominal stress $\mathbf{T}^0$ in terms of the second Piola–Kirchhoff stress $\mathbf{S}$, given here in index form,

$$T_{ij}^0 = S_{ik} F_{kj}^T = S_{ik} F_{jk}$$

Substituting the deformation gradient F_{jk} in terms of the displacement gradient $\partial u_j / \partial X_k$ gives

$$T_{ij}^0 = S_{ij} + S_{ik} u_{j,k} \tag{23.15}$$

where $(\ldots)_{,k}$ denotes $\partial(\ldots)/\partial X_k$. Substituting (23.11) into (23.15) yields

$$T_{ij}^0 = \bar{\sigma}_{ij} + C_{ijkl} \epsilon_{kl} + \bar{\sigma}_{ik} u_{j,k} \tag{23.16}$$

where, again, terms beyond linear in displacement gradient have been neglected. Thus, the change in nominal stress is given by

$$\delta T_{ij}^0 = C_{ijkl}^0 u_{k,l}$$

where

$$C_{ijkl}^0 = C_{ijkl} + \bar{\sigma}_{il} \delta_{kj} \tag{23.17}$$

Because neither the nominal stress T_{ij}^0 nor the displacement gradient $u_{k,l}$ is symmetric, C_{ijkl}^0 is not symmetric with respect to the interchange of the first two and last two subscripts. The moduli (23.17) do, however, satisfy the symmetry $C_{ijkl}^0 = C_{lkji}^0$ as a result of (23.9). Thus, when a component of the pre-stress is comparable to one of the moduli, the difference between C_{ijkl}^0

and C_{ijkl} cannot be neglected. This can occur if either the pre-stress is large or the incremental moduli are small. An example of the first is the interior of the Earth where hydrostatic stress is very large even though strains due to the propagation of waves are small. An example of the second occurs when the response is linearized about a point where the local slope of the stress-strain curve is small.

Equation (23.16) can be rewritten as

$$\delta T_{ij}^0 = C_{ijkl}\epsilon_{kl} + \bar{\sigma}_{ik}\epsilon_{jk} + \bar{\sigma}_{ik}\Omega_{jk} \tag{23.18}$$

where Ω_{jk} is the infinitesimal rotation tensor (17.22) or (17.23). Even if the strains are small and the components of $\bar{\sigma}_{ik}$ are small compared to components of C_{ijkl}, the last term in (23.18) will not be negligible if the product of the pre-stress and the rotation is comparable to the product of the moduli and the strain. A familiar example is the buckling of a column. If κ is the curvature, strains are on the order κh where h is the thickness of the column. The rotations are on the order κl where l is the length of the column. Since buckling typically occurs when $l \gg h$, rotations will be much larger than strains. As a result, buckling is one of the few examples in elementary strength of materials where equilibrium is written for a deformed (slightly buckled) state of the body.

Similar results can be derived for the Cauchy stress σ and the Kirchhoff stress $\tau = J\sigma$ where $J = \det(\mathbf{F})$. Linearizing the expression for the Kirchhoff stress in terms of the second Piola–Kirchhoff stress $\tau = \mathbf{F} \cdot \mathbf{S} \cdot \mathbf{F}^T$ yields

$$\delta\tau_{ij}^* = C_{ijkl}^{\tau}\epsilon_{kl}$$

where

$$\delta\tau_{ij}^* = \delta\tau_{ij} - \Omega_{il}\bar{\sigma}_{lj} - \Omega_{jl}\bar{\sigma}_{li}$$

is the increment of τ_{ij} computed in a frame that is instantaneously rotating with the material and the

$$C_{ijkl}^{\tau} = C_{ijkl} + \frac{1}{2}\{\delta_{ki}\bar{\sigma}_{lj} + \delta_{li}\bar{\sigma}_{kj} + \delta_{kj}\bar{\sigma}_{il} + \delta_{lj}\bar{\sigma}_{ik}\} \tag{23.19}$$

are written symmetrically with respect to the interchange of i and j, k and l, and ij with kl. The incremental moduli for the Cauchy stress are

$$C_{ijkl}^{\sigma} = C_{ijkl}^{\tau} - \bar{\sigma}_{ij}\delta_{kl} \tag{23.20}$$

Note that even if a strain energy density function exists so that $C_{ijkl} = C_{klij}$ and $C_{ijkl}^{\tau} = C_{klij}^{\tau}$, $C_{ijkl}^{\sigma} \neq C_{klij}^{\sigma}$.

We assume that the reference state itself is an equilibrium state,

$$\frac{\partial\bar{\sigma}_{ij}}{\partial X_i} + \rho_0\bar{b}_j^0 = 0 \tag{23.21}$$

where $\rho_0\bar{b}_j^0$ is the body force in the reference state per unit reference volume and the surface traction in the reference state is

$$\bar{t}_j^0 = N_i\bar{\sigma}_{ij} \tag{23.22}$$

Substituting (23.16) into (23.13) and (23.14) and subtracting (23.21) and (23.22) yields

$$\frac{\partial}{\partial X_i}\left\{C_{ijkl}\varepsilon_{kl} + \bar{\sigma}_{ik}\frac{\partial u_j}{\partial X_k}\right\} + \rho_0\left(b_j^0 - \bar{b}_j^0\right) = \rho_0\frac{\partial^2 u_j}{\partial t^2} \tag{23.23}$$

and

$$N_i\{C_{ijkl}\varepsilon_{kl}\} = t_j^0 - \bar{t}_j^0 - N_i\bar{\sigma}_{ik}\frac{\partial u_j}{\partial X_k}$$

where ε_{jk} is the infinitesimal strain from the reference state.

When the terms involving $\bar{\sigma}_{ik}$ can be dropped the usual linear elasticity equations result. This will be the case for the conditions just discussed, but it is worth noting that it is the derivatives of the displacement gradients that enter the equation of motion and these may have magnitudes larger than those of the strains and rotations.

23.2   Linearized Elasticity

Here we specialize immediately to small (infinitesimal) displacement gradients and no pre-stress. This is the conventional formulation of *linear elasticity*. In this case, the stress σ_{ij} is related to the small (infinitesimal) strain tensor by

$$\sigma_{ij} = C_{ijkl}\varepsilon_{kl} \tag{23.24}$$

where C_{ijkl} is an array of material parameters. If the material is *homogeneous*, the material properties are independent of position and the C_{ijkl} are constant. In general, C_{ijkl} has $3^4 = 81$ components but because the stress is symmetric, $\sigma_{ij} = \sigma_{ji}$, as is the strain, $\varepsilon_{kl} = \varepsilon_{lk}$, the number is reduced to $6 \times 6 = 36$. If, in addition, a strain energy density function exists so that the stress is given by

$$\sigma_{ij} = \frac{\partial W}{\partial \varepsilon_{ij}}$$

then C_{ijkl} satisfies the additional symmetry (23.12)

$$C_{ijkl} = C_{klij} \tag{23.25}$$

and the strain energy density is

$$W = \frac{1}{2}\varepsilon_{ij}C_{ijkl}\varepsilon_{kl} \tag{23.26}$$

Because of these symmetries, (23.24) relates six distinct components of stress to six distinct components of strain. Consequently, for an anisotropic material, it is often convenient to treat σ_{ij} and ε_{ij} as six-component vectors that are related by a 6×6 matrix

$$\sigma_i = C_{ij}\varepsilon_j$$

or

$$\begin{bmatrix} \sigma_{11} \\ \sigma_{22} \\ \sigma_{33} \\ \sigma_{23} \\ \sigma_{31} \\ \sigma_{12} \end{bmatrix} = \begin{bmatrix} C_{11} & C_{12} & C_{13} & C_{14} & C_{15} & C_{16} \\ C_{21} & C_{22} & C_{23} & C_{24} & C_{25} & C_{26} \\ C_{31} & C_{32} & C_{33} & C_{34} & C_{35} & C_{36} \\ C_{41} & C_{42} & C_{43} & C_{44} & C_{45} & C_{46} \\ C_{51} & C_{52} & C_{53} & C_{54} & C_{55} & C_{56} \\ C_{61} & C_{62} & C_{63} & C_{64} & C_{65} & C_{66} \end{bmatrix} \begin{bmatrix} \varepsilon_{11} \\ \varepsilon_{22} \\ \varepsilon_{33} \\ 2\varepsilon_{23} \\ 2\varepsilon_{31} \\ 2\varepsilon_{12} \end{bmatrix}$$

where $C_{11} = C_{1111}$, $C_{12} = C_{1122}$, $C_{13} = C_{1133}$, $C_{14} = (C_{1123} + C_{1132})/2$, $C_{15} = (C_{1131} + C_{1113})/2$, $C_{16} = (C_{1112} + C_{1121})/2$, and so on. If a strain energy density function exists, the symmetry (23.25) implies that $C_{ij} = C_{ji}$ and this results in the reduction from 36 to 21 constants for an anisotropic linear elastic material.

23.2.1 Material Symmetry

The number of distinct components of C_{ijkl} can be reduced further if the material possesses any symmetries. Material symmetry can result from crystal structure, processing, or conditions of formation. An example of material processing would be drawing or forming processes. An example of symmetry due to conditions of formation is a sandstone which is formed by deposition in layers.

One approach proceeds along the lines of the discussion of isotropic tensors (6.4). Because C_{ijkl} is a (fourth-order) tensor its components in a coordinate system with unit orthogonal base vectors $\mathbf{e}_i$ must be related to the components C'_{ijkl} in a system of base vectors $\mathbf{e}'_i$ by

$$C'_{ijpq} = A_{ki} A_{lj} A_{mp} A_{nq} C_{klmn}$$

where $A_{ik} = \mathbf{e}'_k \cdot \mathbf{e}_i$. If the material possesses a symmetry such that tests of the material in two coordinate systems cannot distinguish between them, then, for those two coordinate systems, $C'_{ijkl} = C_{ijkl}$ and hence

$$C_{ijpq} = A_{ik} A_{jl} A_{pm} A_{qn} C_{klmn}$$

Suppose, for example, that the $x_1 x_2$ plane is a plane of symmetry. Then a coordinate change that reverses the x_3 axis will not affect the behavior. For such a change, $A_{11} = A_{22} = -A_{33} = 1$ are the only nonzero A_{ij}. Thus

$$C_{1223} = A_{11} A_{22} A_{22} A_{33} C_{1223} = -C_{1223}$$

Hence $C_{1223} = 0$. Similar calculations show that any C_{klmn} having an odd number of threes as indices are zero.

Alternatively, consider the matrix formulation. For changes of coordinate system that are indistinguishable to the material

$$\sigma'_i = C'_{ij} \varepsilon'_i = C_{ij} \varepsilon_j = \sigma_i$$

Again, consider the $x_1 x_2$ plane as a plane of symmetry. Then $\sigma_1' = \sigma_1$ and it follows that

$$
C_{11}\varepsilon_1 + C_{12}\varepsilon_2 + C_{13}\varepsilon_3 + C_{14}\varepsilon_4 + C_{15}\varepsilon_5 + C_{16}\varepsilon_6
$$
$$
= C_{11}'\varepsilon_1' + C_{12}'\varepsilon_2' + C_{13}'\varepsilon_3' + C_{14}'\varepsilon_4' + C_{15}'\varepsilon_5' + C_{16}'\varepsilon_6'
$$

But the shear strains $2\varepsilon_{32} = \varepsilon_4$ and $2\varepsilon_{31} = \varepsilon_5$ reverse sign under the transformation that reverses the x_3 axis; that is, $\varepsilon_4' = -\varepsilon_4$ and $\varepsilon_5' = -\varepsilon_5$. Therefore,

$$
C_{14} = -C_{14}' = -C_{14} = 0
$$

and

$$
C_{15} = -C_{15}' = -C_{15} = 0
$$

The remaining nonzero C_{ij} are

$$
C_{ij} = \begin{bmatrix}
C_{11} & C_{12} & C_{13} & 0 & 0 & C_{16} \\
C_{21} & C_{22} & C_{23} & 0 & 0 & C_{26} \\
C_{31} & C_{32} & C_{33} & 0 & 0 & C_{36} \\
0 & 0 & 0 & C_{44} & C_{45} & 0 \\
0 & 0 & 0 & C_{54} & C_{55} & 0 \\
C_{61} & C_{62} & C_{63} & 0 & 0 & C_{66}
\end{bmatrix}
$$

A crystal structure resulting in a single plane of symmetry is called *monoclinic*.

An *orthotropic* material or a material with orthorhombic crystal structure has symmetry with respect to three orthogonal planes. The nine nonzero C_{ij} are

$$
C_{ij} = \begin{bmatrix}
C_{11} & C_{12} & C_{13} & 0 & 0 & 0 \\
C_{21} & C_{22} & C_{23} & 0 & 0 & 0 \\
C_{31} & C_{32} & C_{33} & 0 & 0 & 0 \\
0 & 0 & 0 & C_{44} & 0 & 0 \\
0 & 0 & 0 & 0 & C_{55} & 0 \\
0 & 0 & 0 & 0 & 0 & C_{66}
\end{bmatrix}
\tag{23.27}
$$

Note that the axial and shear stresses are completely uncoupled.

Hexagonal symmetry is symmetry with respect to 60° rotations. It turns out that this symmetry implies symmetry with respect to any rotation in the plane, which is the same as transverse isotropy. This leaves five nonzero C_{ij}:

$$
C_{ij} = \begin{bmatrix}
C_{11} & C_{12} & C_{13} & 0 & 0 & 0 \\
C_{21} & C_{11} & C_{13} & 0 & 0 & 0 \\
C_{31} & C_{31} & C_{33} & 0 & 0 & 0 \\
0 & 0 & 0 & C_{44} & 0 & 0 \\
0 & 0 & 0 & 0 & C_{44} & 0 \\
0 & 0 & 0 & 0 & 0 & \frac{1}{2}(C_{11} - C_{12})
\end{bmatrix}
\tag{23.28}
$$

Cubic symmetry has three elastic constants. The material has three orthogonal planes of symmetry and is symmetric to 90° rotations about the normals to these planes:

$$C_{ij} = \begin{bmatrix} C_{11} & C_{12} & C_{12} & 0 & 0 & 0 \\ C_{21} & C_{11} & C_{12} & 0 & 0 & 0 \\ C_{21} & C_{21} & C_{11} & 0 & 0 & 0 \\ 0 & 0 & 0 & C_{44} & 0 & 0 \\ 0 & 0 & 0 & 0 & C_{44} & 0 \\ 0 & 0 & 0 & 0 & 0 & C_{44} \end{bmatrix} \tag{23.29}$$

23.2.2   Linear Isotropic Elastic Constitutive Relation

For isotropy, the response of the material is completely independent of direction. This imposes the following additional relation on (23.29):

$$C_{44} = \frac{1}{2}(C_{11} - C_{12}) \tag{23.30}$$

Therefore, a linear elastic isotropic material is described by two elastic constants $C_{12} = \lambda$ and $C_{44} = \mu$. The stress-strain relation is given by

$$\sigma_{ij} = \lambda \varepsilon_{kk} \delta_{ij} + 2\mu \varepsilon_{ij} \tag{23.31}$$

where λ and μ are Lamé constants. If λ and μ are not functions of position, then the material is *homogeneous*.

To invert (23.31) to obtain the strains in terms of the stresses, we first take the trace of (23.31):

$$p = -K\varepsilon_{kk}$$

where $p = -\sigma_{kk}/3$ is the pressure and

$$K = \lambda + \frac{2}{3}\mu$$

is the bulk modulus. Recall that for small displacement gradients ε_{kk} is approximately equal to the volume strain, that is, the change in volume per unit reference volume. Hence K relates the pressure to the volume strain. For an incompressible material $K \rightarrow \infty$; that is, the volume strain is zero, regardless of the pressure. (Note that incompressible does not mean that the material is non-deformable, but only that it deforms with zero volume change.) Substituting for ε_{kk} into (23.31) and rearranging yields

$$2\mu \varepsilon_{ij} = \sigma_{ij} - \sigma_{kk}\delta_{ij}\frac{\lambda}{(3\lambda + 2\mu)} \tag{23.32}$$

Now consider a uniaxial stress: only σ_{11} is nonzero. The strain $\varepsilon_{11} = \sigma_{11}/E$ where

$$E = \frac{\mu(3\lambda + 2\mu)}{(\lambda + \mu)} \tag{23.33}$$

is Young's modulus.

The strain in the lateral direction

$$\varepsilon_{22} = -v\varepsilon_{11}$$

where

$$v = \frac{\lambda}{2(\lambda + \mu)} \tag{23.34}$$

is Poisson's ratio. Equation (23.32) can be rewritten in terms E and v as

$$\varepsilon_{ij} = \frac{(1+v)}{E}\sigma_{ij} - \frac{v}{E}\sigma_{kk}\delta_{ij}$$

Some additional useful relations among the elastic constants are

$$2\mu = \frac{E}{1+v} \tag{23.35}$$

and

$$\lambda = 2\mu\frac{v}{1-2v} \tag{23.36}$$

23.2.3   Restrictions on Elastic Constants

The existence of a strain energy function places restrictions on the values of the elastic constants. These restrictions arise from the requirement that the strain energy function be positive

$$W(\varepsilon) > 0 \tag{23.37}$$

if $\varepsilon \neq 0$ and $W(0) = 0$. The strain-energy density function is given by (23.26). The condition (23.37) requires that C_{ijkl} be positive definite.

For an isotropic material

$$W = \frac{1}{2}\left\{\lambda(\varepsilon_{kk})^2 + 2\mu\varepsilon_{ij}\varepsilon_{ij}\right\} \tag{23.38}$$

Because ε_{ij} and ε_{kk} are not independent, we cannot conclude from (23.37) that the coefficients λ and μ are positive. Consequently, we rewrite (23.38) in terms of the deviatoric strain

$$\varepsilon'_{ij} = \varepsilon_{ij} - \frac{1}{3}\delta_{ij}\varepsilon_{kk}$$

to get

$$W = \frac{1}{2}\left\{\left(\lambda + \frac{2}{3}\mu\right)\varepsilon_{kk}^2 + 2\mu\varepsilon'_{ij}\varepsilon'_{ij}\right\} \tag{23.39}$$

Because each of ε_{kk} and ε'_{ij} can be specified independently, (23.37) requires that the bulk modulus $K = \left(\lambda + \frac{2}{3}\mu\right)$ and shear modulus μ be positive. These conditions require that Young's modulus $E > 0$ and that Poisson's ratio be within the range

$$-1 < \nu < \frac{1}{2} \tag{23.40}$$

Note that for negative ν a bar that increases its length due to uniaxial tension will also increase its cross-sectional area. Although some materials have been fabricated recently with $\nu < 0$, the practical limits on ν are

$$0 < \nu < 0.49$$

Cork is a material with $\nu \approx 0$, a desirable property for use as a stopper in wine bottles. Rubber is nearly incompressible, $\nu = 0.49$. Steel and aluminum have Poisson's ratios of about 0.28 and 0.33, respectively.

23.3   More Linearized Elasticity

The linearized equation of motion with no pre-stress (23.23) is

$$\frac{\partial \sigma_{ij}}{\partial X_i} + \rho_0 b_j^0 = \rho_0 \frac{\partial^2 u_j}{\partial t^2}$$

Substituting (23.31) for an isotropic material, assuming the elastic constants are independent of position (homogeneous), and neglecting the body force yields the Navier equations

$$(\lambda + \mu)\frac{\partial^2 u_k}{\partial X_k \partial X_j} + \mu \frac{\partial^2 u_j}{\partial X_k \partial X_k} = \rho_0 \frac{\partial^2 u_j}{\partial t^2} \tag{23.41}$$

or in coordinate-free notation

$$(\lambda + \mu)\nabla_{\mathbf{X}}(\nabla_{\mathbf{X}} \cdot \mathbf{u}) + \mu \nabla_{\mathbf{X}}^2 \mathbf{u} = \rho_0 \frac{\partial^2 \mathbf{u}}{\partial t^2} \tag{23.42}$$

As a simple example of solutions of this equation, look for displacements of the form

$$u_i = f_i(\mathbf{n} \cdot \mathbf{X} - ct) \tag{23.43}$$

These are solutions for which the displacement is constant on planes with unit normal $\mathbf{n}$ traveling at a speed c. Hence, they represent plane waves. Substituting (23.43) into (23.41) yields

$$(\lambda + \mu)n_j\left(n_k f_k''\right) + \mu f_j'' = \rho_0 c^2 f_j'' \tag{23.44}$$

where $(\ldots)'$ denotes the derivative with respect to the argument. Forming the scalar product of (23.44) with n_j yields

$$c = c_L = \sqrt{(\lambda + 2\mu)/\rho_0}$$

c_L is the bulk or dilatational wave speed. In seismology, this is called the primary wave speed. The displacements are orthogonal to the wave front. Similarly, forming the scalar product of (23.44) with vectors orthogonal to $\mathbf{n}$ yields the shear or transverse wave speed

$$c = c_S = \sqrt{\mu/\rho_0}$$

for which the displacements are parallel to the plane of the wave. Since $c_L > c_S$, the dilatational wave arrives before the shear wave and the difference in arrival times can be used to infer the elastic constants.

23.3.1 Uniqueness of the Static Problem

Solutions to elasticity problems can be obtained in any number of ways. Having obtained a solution, it is important to know that there is no other solution to the same problem. Here we will show that this is the case. Actually, we would like to know that any slight change in the problem formulation leads to solutions that are close in some sense. This can be shown, as well, but is more involved and we will therefore restrict focus to uniqueness.

Thus we will consider a static problem satisfying the equilibrium equation

$$\frac{\partial \sigma_{ij}}{\partial X_i} + \rho_0 b_j^0 = 0 \tag{23.45}$$

the constitutive equation (23.24), the strain displacement equations

$$\varepsilon_{ij} = \frac{1}{2}\left(\frac{\partial u_i}{\partial X_j} + \frac{\partial u_j}{\partial X_i}\right) \tag{23.46}$$

and boundary conditions. The boundary conditions may be of several types:

- Specify the traction $\mathbf{t}$ everywhere on the boundary.
- Specify the displacement $\mathbf{u}$ everywhere on the boundary.
- Specify the traction $\mathbf{t}$ on some portions of the boundary and the displacement $\mathbf{u}$ on all other portions of the boundary.
- Specify one of each pair $t_i u_i$ on some portion of the boundary and either $\mathbf{t}$ or $\mathbf{u}$ on all other portions of the boundary.

There are other possibilities but those above cover most situations. Assume that there are two solutions that satisfy (23.45), (23.24), (23.46), and the boundary conditions and let their difference be denoted by $\Delta u_i, \Delta \sigma_{ij}$. Because both solutions satisfy (23.45)

$$\frac{\partial \Delta \sigma_{ij}}{\partial X_i} = 0 \tag{23.47}$$

Forming the scalar product of Δu_j with (23.47) and integrating over the volume leads to

$$\int_v \Delta u_j \frac{\partial \Delta \sigma_{ij}}{\partial X_i} dv = 0$$

Rewriting the first term as

$$\Delta u_j \frac{\partial \Delta \sigma_{ij}}{\partial X_i} = \frac{\partial (\Delta u_j \Delta \sigma_{ij})}{\partial X_i} - \Delta \sigma_{ij} \frac{\partial \Delta u_j}{\partial X_i}$$

then using the divergence theorem, $\Delta \sigma_{ij} = \Delta \sigma_{ji}$, and (23.46) yields

$$\int_v \Delta \sigma_{ij} \Delta \varepsilon_{ji} \, dv = \int_a \Delta t_i \Delta u_i \, da$$

The right hand side vanishes for any of the boundary conditions specified above. Because both solutions satisfy the same constitutive relation (23.24)

$$\int_v \Delta \varepsilon_{ij} C_{ijkl} \Delta \varepsilon_{kl} \, dv = 0$$

Because of (23.37), C_{ijkl} is positive definite, and the integrand can vanish only if

$$\Delta \varepsilon_{ij} = 0 \tag{23.48}$$

establishing that the stress and strain are unique. The displacements are not unique but (23.48) requires that the difference have the following form:

$$\Delta u_i = A_{ij} X_j + B_i \tag{23.49}$$

where the B_i and A_{ij} are constant and A_{ij} is antisymmetric. Thus, the displacement fields can differ by a translation and rigid rotation. Specifying displacements on a portion of the body rules out this possibility.

23.3.2   Pressurized Hollow Sphere

As an example of the solution of a linear elastic boundary value problem consider the hollow sphere subjected to internal and external pressure shown in Figure 23.1. The sphere is subjected to a pressure P_b at radius $r = b$ and a pressure P_a at radius $r = a$. Obviously the displacement depends only on the radial coordinate $r = \sqrt{X_k X_k}$. This problem is, perhaps, more naturally solved in spherical coordinates. Nevertheless, solution in rectangular Cartesian coordinates presents little additional complications by noting that the components of a unit vector in the radial direction are X_i/r. Consequently, the Cartesian components of the displacement must have the form

$$u_i = \frac{X_i}{r} u(r) \tag{23.50}$$

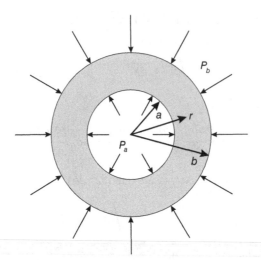

Figure 23.1   Internally and externally pressurized sphere.

where $u(r)$ is the displacement in the radial direction. Because we have focused on rectangular coordinates we will use this approach.

The boundary condition at $r = b$ is

$$(X_i/b)\sigma_{ij} = -P_b(X_j/b) \tag{23.51}$$

and at $r = a$ is

$$-(X_i/a)\sigma_{ij} = P_a(X_j/a) \tag{23.52}$$

where the minus sign in (23.51) occurs because the pressure is in the negative radial direction and that in (23.52) because the normal to the boundary is in the negative radial direction. The derivatives of (23.50) needed for substitution in (23.41) are calculated as follows:

$$\frac{\partial u_i}{\partial X_j} = u(r)\left\{ \frac{\delta_{ij}}{r} - \frac{X_i X_j}{r^3} \right\} + \frac{X_i X_j}{r^2} u'(r)$$

where $u'(r) = du/dr$. Calculating the other derivatives in similar fashion and substituting in (23.41) with zero right side gives

$$(\lambda + 2\mu)\frac{X_i}{r}\{r^2 u''(r) + 2ru'(r) - 2u(r)\} = 0$$

and, consequently, the term $\{\ldots\}$ must vanish. Looking for a solution of the form r^n reveals that

$$u(r) = \frac{A}{r^2} + Br$$

where A and B are constants. The first term gives a strain that is purely deviatoric and the second a strain that is a uniform dilatation.

The stress components calculated from (23.31) are

$$\sigma_{ij} = B\delta_{ij} + \frac{A}{r^3}\left\{\delta_{ij} - \frac{3X_iX_j}{r^2}\right\} \tag{23.53}$$

where A and B have been redefined to absorb the elastic constants. The radial component of stress, σ_r, is

$$\sigma_r = \frac{x_i\sigma_{ij}x_j}{r^2} = B - \frac{2A}{r^3}$$

Substituting into (23.51) and (23.52) and solving for A and B gives

$$A = (P_a - P_b)\frac{a^3b^3}{2(b^3 - a^3)}$$

and

$$B = \frac{\left(P_aa^3 - P_bb^3\right)}{b^3 - a^3}$$

Because A and B are independent of the elastic constants, so is the stress field, a consequence of the all-traction boundary conditions. The hoop stress σ_θ can be calculated by taking the trace of (23.53), noting that it is invariant and must equal $\sigma_r + 2\sigma_\theta$ in spherical coordinates.

Exercises

23.1 Derive (23.9) directly from (23.8) by regarding the strain energy as a function of the deformation gradient, i.e., $W = W(\mathbf{F})$, and computing

$$\frac{\partial W}{\partial F_{mn}} = \frac{\partial W}{\partial E_{ij}}\frac{\partial E_{ij}}{\partial F_{mn}}$$

23.2 The constitutive equation for an elastic material with a strain energy density function W can be expressed as (23.8) where S_{ij} is the second Piola–Kirchhoff stress and E_{ij} is the Green–Lagrange strain. For an isotropic material W can be expressed in terms of the invariants of the Green strain or, equivalently, the invariants of the Cauchy deformation tensor $\mathbf{C} = \mathbf{F}^T \cdot \mathbf{F}$. Thus, $W = W(I_1, I_2, I_3)$ where the invariants I_1 and I_2 are given by (7.9) and (7.10) and I_3 is given by the result of Problem 7.3.
(a) Show that the second Piola–Kirchhoff stress $\mathbf{S}$ is given by

$$\mathbf{S} = 2\{W_1 - I_1W_2 - I_2W_3\}\mathbf{I} + 2\{W_2 - I_1W_3\}\mathbf{C} + 2W_3\mathbf{C}\cdot\mathbf{C}$$

where $W_i = \partial W/\partial I_i$.

(b) By using the relation $\sigma = |\mathbf{F}|^{-1}\mathbf{F} \cdot \mathbf{S} \cdot \mathbf{F}^T$, show that the Cauchy stress σ is given by

$$\sigma = 2|\mathbf{B}|^{-1/2}\{I_3 W_3 \mathbf{I} + (W_1 - I_1 W_2)\mathbf{B} + W_2 \mathbf{B} \cdot \mathbf{B}\}$$

where $\mathbf{B} = \mathbf{F} \cdot \mathbf{F}^T$ is the Finger deformation tensor.

(c) Use the result of Problem 7.4.a in (b) to show that

$$\sigma = 2|\mathbf{B}|^{-1/2}\{(I_3 W_3 + I_2 W_2)\mathbf{I} + W_1 \mathbf{B} + W_2 I_3 \mathbf{B}^{-1}\}$$

23.3 Derive (23.19) and (23.20).

23.4 Show that the difference between the linearized versions of the Cauchy stress and the Kirchhoff stress is negligible if the volume strain is small.

23.5 A constitutive relation for heat conduction is Fourier's law

$$q_i = K_{ij}\frac{\partial T}{\partial x_j}$$

where $\mathbf{q}$ is the heat flux vector, T is the temperature, and $\mathbf{K}$ is the thermal conductivity tensor, a material property.

(a) If the heat flux can be expressed in terms of a scalar potential function G,

$$q_i = \frac{\partial G}{\partial T_{,i}}$$

where $T_{,i} = \partial T/\partial x_i$, what condition does this impose on $\mathbf{K}$, the thermal conductivity tensor? In this case, what is the number of independent components of $\mathbf{K}$?

(b) If the $x_1 x_3$ plane is a plane of material symmetry, determine the reduced form of the conductivity tensor.

(c) If, in addition, the $x_1 x_2$ plane is a plane of material symmetry, again determine the reduced form of the conductivity tensor.

(d) If the material is isotropic, determine the form of $\mathbf{K}$.

23.6 Show that material symmetry with respect to 60° rotations about the X_3 axis reduces (23.27) to (23.28). Verify that such a material is symmetric with respect to any rotation about the X_3 axis.

23.7 Show that an isotropic material imposes the additional relation (23.30) on (23.29).

23.8 Derive (23.33) and (23.34).

23.9 Derive (23.35) and (23.36).

23.10 Determine the modulus M for uniaxial strain

$$\sigma_{11} = M\varepsilon_{11}$$

where ε_{11} is the only nonzero strain component and determine the ratio σ_{22}/σ_{11}.

23.11 A state of plane strain exists if $\varepsilon_{33} = \varepsilon_{13} = \varepsilon_{23} = 0$.
    (a) Show that for plane strain

$$\sigma_{33} = v(\sigma_{11} + \sigma_{22})$$

    (b) Show that

$$\varepsilon_{\alpha\beta} = \frac{1+v}{E}\{\sigma_{\alpha\beta} - v(\sigma_{11} + \sigma_{22})\}$$

    where $\alpha, \beta = 1, 2$.

23.12 Show that for an isotropic material the strain energy can be written as (23.39).

23.13 Show that the conditions $K > 0$ and $\mu > 0$ require $E > 0$ and (23.40).

23.14 In the solution for the pressurized sphere take $b \to \infty$, $P_b \to 0$, and $a \to \infty$ but maintain the product $a^3 P_a = m$ as finite. Specialize the stress and displacement fields for this case. This is the singular (since the stress and displacement become unbounded) solution of a *center of dilatation*.

23.15 The spherically symmetric solution can also be used to solve the problem of a spherical inclusion, a special case of a more general solution by Eshelby (1957). Consider an infinite material in which a spherical region of radius $r = a$ undergoes a *transformation strain* which would correspond to an increase of radius to $a + \varepsilon_0 a$ in the absence of the constraint of the surrounding material. This transformation is assumed not to change the elastic properties of the region. Depending on the application, this strain may be due to a phase transformation, injection of fluid, increase of temperature, etc. It is desired to determine the actual strain undergone in this region in the presence of the constraint of the surrounding material. Eshelby (1957) solved this problem by an ingenious procedure of cutting, transforming, and reinserting the inclusion. The procedure is outlined in this problem.
    (a) Remove the spherical region of radius a from the material. Due to phase transformation etc., the region increases its radius to $a + \varepsilon_0 a$ *without stress*. Show that ε_0 is one-third of the volume strain.
    (b) Determine the pressure p_a that must be applied at $r = a$ to restore the radius of this region to a. Because the inclusion is its original size, it can be reinserted without causing stress. But the material now contains a force layer at $r = a$ corresponding to

$$\sigma_{rr}^-(a) = \sigma_{rr}^+(a) + p_a$$

    where the plus and minus signs indicate that boundary is approached from r greater than or less than a. Using the appropriate solutions inside and outside of a, the condition above, and continuity of displacements show that the actual strain of the inclusion (region $r \le a$) is

$$\varepsilon = \frac{\lambda + 2\mu/3}{\lambda + 2\mu}\varepsilon_0$$

    and that the pressure in the inclusion is $4\mu\varepsilon$.

23.16 Consider the same problem but take the point of view that transformation of the inclusion causes a spherical dislocation, that is, a radial displacement discontinuity

$$u^+ = u^- + \varepsilon_0$$

Show that using this condition and continuity of radial stress yields the same result as the preceding problem.

Reference

Eshelby JD 1957 The determination of the elastic field of an ellipsoidal inclusion and related problems. *Proceedings of the Royal Society of London A* **241**, 376–396.

Index

Fundamentals of Continuum Mechanics, First Edition. John W. Rudnicki.
© 2015 John Wiley & Sons, Ltd. Published 2015 by John Wiley & Sons, Ltd.

Printed in the United States
By Bookmasters